Safety Practices, Firm Culture, and Workplace Injuries

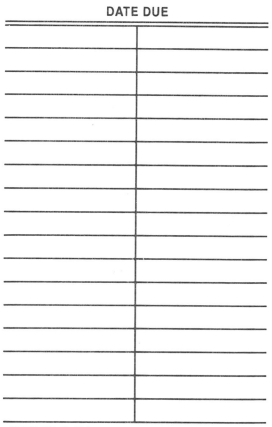

DATE DUE

Safety Practices, Firm Culture, and Workplace Injuries

Richard J. Butler
Yong-Seung Park

2005

W.E. Upjohn Institute for Employment Research
Kalamazoo, Michigan

Library of Congress Cataloging-in-Publication Data

Butler, Richard J.
 Safety practices, firm culture, and workplace injuries / Richard J. Butler, Yong-Seung Park.
  p. cm.
 Includes bibliographical references and index.
 ISBN-10: 0-88099-275-1 (pbk. : alk. paper)
 ISBN-13: 978-0-88099-275-6 (pbk. : alk. paper)
 ISBN-10: 0-88099-277-8 (hardcover : alk. paper)
 ISBN-13: 978-0-88099-277-0 (hardcover : alk. paper)
 1. Industrial safety—Management. 2. Accidents—Prevention. I. Park, Yong-Seung.
II. Title.
 T55.B87 2005
 658.3'82—dc22

                          2005004401

The facts presented in this study and the observations and viewpoints expressed are the sole responsibility of the authors. They do not necessarily represent positions of the W.E. Upjohn Institute for Employment Research.

Cover design by Alcorn Publication Design.
Index prepared by Diane Worden.
Printed in the United States of America.
Printed on recycled paper.

Contents

Figure

Tables

Acknowledgments

We are grateful to the W.E. Upjohn Institute for Employment Research in Kalamazoo, Michigan, both for its staff's many insightful suggestions, which improved the quality of the final version of this book, and for its financial support, which made this project possible. We are particularly grateful to H. Allan Hunt, assistant director of the Institute, whose comments on the first draft encouraged major revisions and a more integrated approach to the literature review; to Erika Jackson, the book's typesetter; and to Benjamin Jones, our conscientious editor at Upjohn.

During the initial course of this research, we also benefited from the suggestions of several colleagues at the Industrial Relations Center, Carlson School of Management, University of Minnesota, in Minneapolis. In this regard, we particularly benefited from the research of Avner Ben-Ner, which shaped much of our research design. We also profited from the generous support and helpful comments of the staff at the Minnesota Department of Labor and Industry in St. Paul, without whose survey and research expertise this project would not have been possible. The final revisions of this book benefited from support provided by the Department of Economics at Brigham Young University in Provo, Utah.

Finally, our deepest gratitude goes to our wives and families for their tolerance and love.

1

Human Resource Management and Safety

Technical Efficiency and Economic Incentives

More U.S. workers die each year on the job than were killed in the U.S. military cumulatively from 1998 through November 2004, even after including self-inflicted and accidental military deaths (DIOR 2005). In 2001, there were 8,786 job-related fatal injuries (5,900 not counting the fatalities caused by the terrorist attacks of September 11), or about 3.7 fatal injuries per 100,000 workers. Workers made 2.1 million trips to the emergency room for injuries sustained from accidents at work (Centers for Disease Control and Prevention 2004). Workers' compensation insurance, which covers all medical expenses and part of lost wages associated with injuries, cost employers $63.9 billion in 2001 (Williams, Reno, and Burton 2003). The indirect costs of accidents—lost wages, damage to equipment, and training and rehabilitation expenses—were several times this amount.

Human resource management (HRM) is usually viewed as an auxiliary function in a firm, contributing nothing to that firm's output—a cost tolerated because payroll, benefits, and certain types of human resource activity must be organized before the real job of production can be undertaken. But HRM practices can affect accident costs in three ways. Two of the three pertain to the real or intrinsic risk in the workplace. "Real" risk is the level of physical danger of accidental injury or occupational disease that comes from workers producing output. As men interact with machines, both men and machines cause accidents. Accidents can be reduced by modifying either part of the interaction: 1) by increasing workers' incentives to be careful, or 2) by modifying the workplace environment to employ processes, procedures, equipment, and ergonomics that reduce on-the-job injuries. In addition, HRM policy can reduce accident costs by lessening workers' incentives to file false or inflated accident claims for any given level of real risk.

1

Most models of firm behavior ignore HRM, assuming management simply chooses labor and capital to maximize firms' profits. As the product price, the wage rate, or the rental cost of machinery changes, so do the optimal production of widgets and the optimal configuration of inputs. In these traditional models, labor is passive with respect to the production process in two ways: labor does nothing to improve the technical efficiency of the firm, and labor always acts in the firm's best interest, regardless of labor's own incentives. Management is assumed to know everything the workers know.

THE IMBALANCE CREATED BY ASYMMETRIC INFORMATION

But the traditional model is seriously flawed, as there is ample evidence that workers are not passive. What workers know about their own behavior or about the firm's technology may profoundly affect profitability. Costs may also rise because of excessive consumption of fringe benefits by employees when legitimate claims are difficult to distinguish from questionable claims. Or management may ignorantly be providing inferior plant design or unsafe production processes—resource misallocation that could be improved with labor's help. Such asymmetric information, where employees know something that it is difficult or costly for management to know, can yield costs that are unnecessarily high.

The collapse of the Enron Corporation is an example of how important asymmetric information can be, though the information asymmetry there was largely between management and shareholders. Enron managers and accountants deceived shareholders into believing the company was in much better financial shape than it actually was, having information about company debt and revenue that the general public did not have. This asymmetry of information was exploited by management, inflating company stock value beyond its actual worth in order to increase management income and maintain management control.

While asymmetric information problems between management and shareholders, such as happened at Enron, are a spectacular type of asymmetric information problem, such problems also exist between management and company employees. In most jobs it is impractical to monitor all employee behavior. So employees know more about their work effort and level of care than managers do. Where such asymmetric information exists, questions arise: Are employees working as hard as they have agreed to work under the employment contract? Are the employees treating company property with the same respect and due care with which they would treat their own property, so that there is no thievery, vandalism, or misuse of company equipment? Are the employees being as safe as prudently possible? Are the employees only using sick days when they need to use them? Are the employees only filing lost-work insurance claims for legitimate, on-the-job injuries? These are all areas where asymmetric information can drive a wedge between what management expects and what employees deliver.

HRM Practices Can Treat Asymmetric Information Problems

HRM practices are increasingly viewed as one way the firm can address asymmetric information problems. The hypothesized causal link between HRM practices and a reduction in asymmetric information problems has HRM changing profits: HRM programs provide workers with incentives to change their behavior by aligning their activities with management's profit objective. Without these incentives, profits are lower.

This model cannot be tested in its entirety; too many model components remain either unmeasured or unmeasurable. For example, asymmetric information is not public information; it is not measured. Profits and even costs are not uniformly reported for all companies, especially for small and medium-sized companies such as we have in our sample. However, there is one important category of cost—safety costs—which is measured in sufficient detail to use in testing our model. The test is simple: do alternative HRM practices affect employees' injury claims? Do some HRM practices help reduce injury-claim frequency? Do other HRM practices help reduce injury-claim severity? If they do reduce safety costs, is it because the HRM practices are improving technical efficiency or because HRM practices are reducing disability benefits consumption associated with asymmetric information?

In this book, we estimate how various HRM practices affect occupational safety: which HRM practices lower firms' workers' compensation costs and whether their impact comes through changes in technical efficiency or through induced changes in workers' behavior. We present a model of safety outcomes in this chapter that illuminates the ways in which HRM might affect safety outcomes, and in the next chapter we use this model as a basis for reviewing the empirical research in this field. We present our own research findings in Chapters 3 and 4. In Chapter 5 we draw conclusions.

THE ACCIDENTS-OUTPUT TRADEOFF

Either improved technical efficiency or labor incentives will lower firms' overall safety costs, including outlays for machinery, compensating differentials for risk, and lost work time. Frequently we ignore the overall influence of safety. One long-recognized shortcoming of the simple classical model is that it fails to take into account that the firm produces not only output but accident claims with a given level of labor and capital inputs. That is to say, in real-world industrial processes, accidents are a natural by-product of production. Getting rid of all job accidents is often feasible only with very large reductions in output. Even white-collar workers occasionally bump their heads in their cubicles or get paper cuts that may become infected; wearing special headgear and thick mittens could prevent such accidents. We rarely wear headgear or mittens in the office because the reduction in productivity would outweigh any gains in safety. On the other hand, sometimes we can increase the level of output by ignoring prudent safety precautions—such as by taking off the safety guards from machinery or removing the guardrails along catwalks and stairs that inhibit the movement of materials—but accident costs resulting from ignoring basic safety practices generally outweigh the additional output that would be garnered by doing so. There is a tradeoff between accidents and output.[1]

Another shortcoming of the classical model, mentioned earlier, is that the human factors in the production process are not passive—employees may react to incentives, and employees may provide valuable information about the optimal organization of production. For example, when construction workers show management how a wall can be framed more safely and quickly by assembling it horizontally on the ground, rather than piecing it together vertically in the air, they provide valuable information on the technical efficiency of the process. Because workers are assembling products every day and building techniques are constantly changing, opportunities for uncovering such technical efficiencies are abundant. When HRM practices lead to improved techniques of production, technical efficiency improves. Accident prevention costs, broadly defined, fall.

HRM Practices Can Also Treat Safety Behavior Problems

Likewise, variations in HRM practices may change worker safety behavior and the firm's safety costs. Consider the time path of injury benefits, shown in Figure 1.1, typical of workers' compensation laws in most states—including Minnesota, from which we draw our sample for this study.

Figure 1.1 Time Path of Injury Benefits

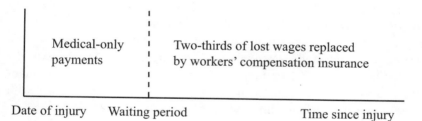

The first three days following the injury are known as the waiting period in workers' compensation. (It is three days in Minnesota; other common waiting periods are five and seven days.) During the waiting period, the injured worker receives no lost wage benefits, though all of the injury-related medical costs are covered by the firm's workers' compensation insurance policy. Hence, during the waiting period there are only payments for medical treatments. After the waiting period, two-thirds of the worker's lost wages are replaced by indemnity payments (payments for lost wages, in addition to the medical payments). Both the waiting period and the partial wage replacement are types of insurance cost sharing. Insurance contracts are structured so that whenever a worker is injured, he bears part of the wage-loss risk.

THE PERILS OF MORAL HAZARD

Cost sharing, such as waiting periods and partial wage replacement, mitigates incentive problems under asymmetric information. Behavioral changes resulting from incentives generated by disability and health insurance coverage are known as *moral hazard*. In other words, moral hazard exists when workers (or firms or health care providers) change behavior for personal gain under an insurance contract.

There are many actions an insured worker or health care provider can take that affect the size or the probability of a loss.[2] With respect to injuries that are temporarily disabling, for example, an insurance contract may specify that the only disabilities covered are those arising from injuries sustained while on the job. Hence, a worker may claim a given condition arose from a job injury and may seek temporary total disability benefits because his real health condition does not qualify under the contract. Or, the worker may have a recurring health condition, such as lower back pain, which in the absence of insurance he simply tolerates, since treatment would impose personal costs. When insured, however, he may choose not to work and incur health service costs and draw disability benefits since others are paying the benefits. An extreme case of behavioral change might be overt fraud in which a worker—facing a pending layoff—claims injury benefits when no injury or health condition was incurred, either on or off the job.

Insurance contracts recognize that moral hazard is costly. If insurers or firms had full information about all workplace injuries, they could reimburse workers for all lost wages until those workers returned to work. Under full information, the firm would know what injuries were work-related and the firm would know exactly when the worker was able to return to work. But firms don't usually have such information. Monitoring the behavior of all participants in an insurance contract is costly, and the costs of such monitoring generally exceed the benefits. The workers know this, and since they have considerable latitude in changing their behavior to enhance their short-run well-being, they sometimes behave differently than they would in the absence of insurance payments. Therefore, the root of the moral hazard problem is an information asymmetry between workers and firms—workers know more about their own health status, as well as their preference for leisure over work (hence their willingness to feign injury if it results in paid leisure time), than firms know.

Workers' compensation insurance recognizes the complex incentives generated by disability coverage and alters the benefits contract so workers bear some of the costs of the injury themselves. They get no lost-wage pay during the waiting period and only partial reimbursement thereafter. These cost-sharing arrangements exist to induce workers to take an appropriate amount of care on the work site and only the necessary time off work. HRM practices can affect those incentives.

An Illustrative Model of Three Cases of Worker Reimbursement

Consider a small construction company (Table 1.1A,B,C). Initially assume the firm's policy is to allow sick-day pay to reimburse workers for lost wages during the waiting period. We will call this the usual case. We also assume a simple fixed input-output process (a Leontief technology) that requires five laborers to construct a building.[3] Only construction workers actually build; the foremen supervise workers and help replace the labor services of injured construction workers. We assume the rate of building depends only on the number of construction laborers, but the rate of concomitant accidents varies with the degree of care exercised by the workers and the supervising foremen.

Since output is fixed, the firm's economic problem is to minimize the sum of labor costs and safety costs. In this example, each foreman is paid $100,000 and each construction laborer is paid $40,000. Each accident costs $30,000 in terms of replacement labor and capital costs. These are the only costs associated with on-the-job accidents. Initially, suppose a workers' compensation system is in place that only pays some of the lost wages after the waiting period, though the firm's HRM practices allow workers to use their sick-day benefits to replace their lost wages for the first three days following an injury. Hence, injured workers bear some costs of workplace injuries, though not any costs associated with the waiting period.

Table 1.1A is the usual case, before any changes in standard HRM practices are implemented. Our assumed Leontief technology is such that with the number of laborers fixed, the output is fixed, and there is no substitution between foremen and laborers in building production. While adding more foremen doesn't increase the number of buildings,

Table 1.1A Usual Case: Partial Wage Replacement after the Waiting Period

Number of foremen	Number of laborers	Accidents	Wage/salary costs ($)	Accident costs ($)	Total costs ($)
1	5	8	300,000	240,000	540,000
2	5	4	400,000	120,000	520,000[a]
3	5	1	500,000	30,000	530,000
5	5	0	700,000	0	700,000

[a] Optimal cost-benefit level.

it does lower the number of accidents. Foremen monitor the safety content of work, and with more foremen present, safety costs fall, although at a diminishing rate. Since output is fixed, the firm maximizes its profits by minimizing the costs.

The tradeoff in Table 1.1A is simple: more foremen reduce accident costs but increase wage costs. The firm's optimal allocation rule will be to add foremen until the marginal cost of the additional foremen (in terms of the increase in wage costs) is greater than the marginal benefit of the additional foremen (in terms of the reduction in safety costs). In Table 1.1A, the cost-minimizing level of output is produced by going with two foremen. Going from one foreman to two increases the wage costs by $100,000 while it reduces the number of accidents from eight to four, saving $120,000 in accident costs. However, going from two foremen to three increases overall costs: wage costs rise by $100,000 while accident costs only fall by $90,000.

Even though the firm could construct its buildings without any accidents by hiring five foremen, it does not choose to do so. The additional costs (in terms of foremen's wages) do not justify the additional gains from producing with no injuries. It is not optimal to reduce the injuries to zero. Indeed, in each of the three cases we examine in Table 1.1, it is cheaper to allow some injuries than it is to do away with all injuries.

HRM Practices Can Worsen Incentive or Moral Hazard Problems

Suppose that we change HRM policy, but in a way that provides fewer incentives for workers to take care. Specifically, suppose that the new HRM policy guarantees that all lost wages due to an injury will be reimbursed, not just those of the initial waiting period, without time limitation or a financial cap to the total benefits received: the company makes up any difference between the employees' wages and workers' compensation benefits through a wage continuation policy that guarantees that 100 percent of the worker's nominal wage will be replaced. With their pay as high on a workers' compensation claim as it is on the job, 1) workers could take more risks on the job than they formerly did when they bore some of the wage costs of an accident, changing their real safety behavior, or 2) workers might simply report more accidents than they formerly did, given the same level of risk. The former is called risk-bearing moral hazard; the latter, claims-reporting moral hazard (Butler and Worrall 1991). As the workers' insurance coverage under the new HRM policy expands, moral hazard potential increases and the number of reported claims rises.

An example of this rise in injury rates, holding the level of monitoring constant, is given in Table 1.1B. While the output remains constant at the same level it did in Table 1.1A, the accidents double for each combination of laborers and foremen. With one foreman and five laborers, the number of reported accidents goes from eight with normal care to 16 when laborers take less care because of moral hazard response. As the number of reported injuries doubles, the value of additional foremen increases. In Table 1.1A, going from one to two foremen decreases accident costs by $120,000; in Table 1.1B, going from one to two foremen decreases accident costs by $240,000. Because the marginal cost of foremen stays constant, the increased marginal benefit of additional foremen increases the firm's demand for their monitoring activity, and the optimal number of foremen rises. In Table 1.1B the potential for moral hazard behavior has increased, and the optimal number of foremen has risen from two to three.

Finally, as a direct result of the increase in the moral hazard under the new HRM policy reflected in Table 1.1B, there are more claims, so safety costs are higher for every combination of input (except for where there are five foremen; here the costs remain zero).

HRM Practices Can Also Improve Incentive or Moral Hazard Problems

Suppose that instead of "topping off" disability benefits so there were no wages lost when workers were injured, the firm adopted a

Table 1.1B   Moral Hazard Response: Full Wage Replacement Benefits Lower Workers' Incentive to Take Care

Number of foremen	Number of laborers	Accidents	Wage/salary costs ($)	Accident costs ($)	Total costs ($)
1	5	16	300,000	480,000	780,000
2	5	8	400,000	240,000	640,000
3	5	2	500,000	60,000	560,000[a]
5	5	0	700,000	0	700,000

[a] Optimal cost-benefit level.

policy that moved in a different direction: it adopted the HRM practices of Table 1.1A with respect to disability benefits (only two-thirds of the wage is replaced following an injury), but now it has added a profit-sharing plan in which it distributes 10 percent of the company's profits to workers. Especially for a small company where workers can more readily monitor each other and apply peer pressure (so there is less likely to be a free rider response that mitigates the financial incentives), this is likely to align employees' incentives with management's profit-maximizing efforts.[4] We may suppose employees respond to such profit-sharing incentives by being more careful on the job or simply by filing fewer claims than in Tables 1.1A and 1.1B.

In Table 1.1C, the input combinations are the same as those in Tables 1.1A and 1.1B, but there are fewer accidents that result at each level of input: Table 1.1C input combinations now have only half the accident rates of Table 1.1A, and only one-fourth the accident rates of Table 1.1B. Total costs are naturally lower for each combination of inputs, and marginal benefits of monitoring are lowered as well. Going from one foreman to two reduces accident costs by only $60,000, but it costs $100,000 in additional salary to obtain this reduction: the marginal benefits from safety monitoring have fallen, but the marginal costs stayed the same. Hence, less monitoring is optimal and only one foreman will be hired to work with the five laborers. If we assume that company revenue is $520,000, this implies profit sharing of $10,000 with one foreman (10 percent of profits = [$520,000 − $420,000] × 0.1), $6,000 with two foremen, and $500 with three foremen. This minimizes total costs at $420,000, given the new worker incentives induced by the HRM changes. Indeed, in the absence of profit sharing, the firm would revert to Table 1.1A outcomes and total costs would increase by $90,000 (accounting for the profit-sharing payout).

HRM Practices Can Improve or Worsen Technical Efficacy

While the examples in Table 1.1 focused on workers' incentives (through risk-bearing and claims-reporting moral hazard) as HRM practices changed, those tables could just as well have represented changes in production efficiency (through physical ergonomic changes) induced by changes in HRM practices. If a change in HRM practices discouraged communications between worker and firm, it could worsen technical efficiency, and the results could be those pictured in Table 1.1B. For example, if HRM practices included safety standards that didn't improve safety but limited productivity—say, wearing thin, slippery silk gloves when handling power equipment in an effort to reduce carpal tunnel syndrome—then the change in HRM practices could conceivably increase accident costs. On the other hand, if the implementation of new HRM practices improves communications between the worker and the firm in a way that results in fewer accidents for each level of output, then costs would tend to change as they did in Table 1.1C. Assembling some components on the ground and then hauling them into place might be one such improvement. Changes in assembly sequencing, tool usage, and even product design might be other such improvements.

Table 1.1C Profit Sharing Initiated: Incentive Rises to Behave So as to Maximize Profits

Number of foremen	Number of laborers	Accidents	Wage/salary costs ($)	Accident costs ($)	Total costs before profit sharing ($)
1	5	4	300,000	120,000	420,000[a]
2	5	2	400,000	60,000	460,000
3	5	0.5[b]	500,000	15,000	515,000
5	5	0	700,000	0	700,000

[a] Optimal cost-benefit level. Assuming total revenue is $520,000, total costs including profit sharing are $430,000, $466,000, $515,500, and $700,000, depending on number of foremen.

[b] Represents one accident every other period.

The discussion of Table 1.1 illustrates the issues addressed in this book: the extent to which changes in HRM practices change accident costs, which HRM practices are most effective, and whether those result in moral hazard changes or changes in technical efficiency. In a world of perfect certainty and full information, the firm would always adopt those HRM practices that were optimal, producing the best combination of technical efficiency and economic incentives. So any expansion of a practice, or adoption of a new practice, would result in lower accident costs, given that output was held constant. But the optimal combination might not always be obvious to firms because of informational asymmetries, contract restrictions, poor management incentives, or inept bureaucratic procedures. In this study, we address these issues using a sample of Minnesota firms. We chose firms from that state because we have extensive data on their workers' compensation costs as well as their HRM practices. Using this sample, we hope to estimate not only which HRM practices are most cost-effective, but also whether they reduce costs through a reduction in moral hazard or an increase in technical efficiency.

Notes

1. See Walter Oi (1974) for an extensive analysis of this tradeoff.
2. See Butler, Gardner, and Gardner (1997) for empirical examples and citations to the empirical safety literature.
3. Wassily Leontief, a Nobel laureate in economics, pioneered the use of production functions where the ratios of inputs to outputs were fixed so there was no substitution between inputs. This type of production has been used extensively in short-term business forecasting and production planning.
4. The free rider problem arises when one worker does not incur the costs of taking care, thinking that all other workers will take care instead. Thus he is a "free rider" in that he doesn't incur the costs but plans to enjoy the benefits (the extra profits) generated when others take care.

2
Prior Studies of Human Resource Management and Safety

The description at the end of Chapter 1 makes it clear that not all human resource management (HRM) changes necessarily improve safety. HRM policies implemented to eliminate all cost sharing associated with injuries, while they do remove income risk for workers, also lessen their safety incentive. An example would be policies that ensured that employees on workers' compensation claims made as much money as they would at work. Profit sharing, on the other hand, might increase employees' safety awareness. In this chapter, we review prior analyses of how various HRM practices affect safety to get a better idea of which practices improve safety and which do not.

The workplace environment affects safety not only directly, through the provision of safety procedures and ergonomic equipment, but also indirectly, through interpersonal relationships and incentives generated by human resource policy. Certainly, work can be stressful, and this can harm workers' health. Since the early days of scientific work management (Taylor 1947), workers have reported job dissatisfaction, psychological stress, and injuries (Walker and Guest 1952). As office work becomes automated, health problems and work-related diseases continue to be important issues in industrial relations and human resources (Smith et al. 1981). Some researchers believe that stress results from new organizational practices: loss of task control at work, scheduling demands, greater specialization, increased electronic monitoring, and job insecurity (Cooper and Smith 1985; Cyert and Mowery 1988; Majchrzak 1988; Smith et al. 1981, 1992). Research has accumulated on the relationship between the work environment and stress-related diseases such as mental health and heart disease (Cooper and Marshall 1976; House 1981; Karasek 1979; Kasl 1978; Smith 1981, 1987). Recently there has been increased interest in the effect of workplace organization on the incidence of cumulative trauma disease (Moon and Sauter 1996).

These modern trends—and the press of domestic and international competition—force companies to adopt new technologies, attempt new management practices, and make use of new models of work organization (Appelbaum and Batt 1994; Ben-Ner and Jones 1995; Cappelli et al. 1997; Levine and Tyson 1990). Prominent among the changes are human resource management (HRM) practices that provide employees with participation rights in the decision-making process and financial ownership rights in the firm. While these high-performance HRM practices have attracted the attention of practitioners and researchers, previous research has largely been limited to how work organization practices affect firm productivity and profitability. Little is known about how the new HRM practices affect work safety; only a few studies have estimated how management culture influences workplace accident costs.[1] The research reported in this study examines all three dimensions of this HRM revolution: worker involvement in decision making, worker involvement in financial returns, and management involvement in the firm's safety process (i.e., management safety culture).

EMPLOYEE PARTICIPATION IN DECISION MAKING

Employee participation in decision making may increase information about optimal policies. Granting employees the right to participate in the firm's strategic safety policy—for example, increasing worker involvement in the design and implementation of safety policy—may improve workplace safety for both informational and psychological reasons. Employee involvement in strategic planning usually occurs on company time, so employees are essentially acting as paid consultants. In this role, workers may cost-effectively identify safety improvements. Workers are more intimately involved with risky production processes—including other workers' responses to those processes—and so can identify and monitor risk at a lower cost than firm managers or outside consultants (Eaton and Nocerino 2000; Shannon et al. 1996). Involvement with the introduction of new production technologies allows workers to voice their concerns about risk exposure during the decision-making process. This may reduce turnover, increase worker morale, and maintain the firm's stock of specific human capital—all of which may enhance safety outcomes.

A second benefit from employee involvement is that workers pick up information about workplace risk. Workers' getting more information concerning job risk and the firm's safety efforts will help reduce uncertainty in their minds and thus increase the value of the employment experience for risk-averse workers. This lowers employment costs and may improve worker productivity to the extent that safety behaviors depend upon workers' attitudes and improved morale leads to higher productivity.

Another psychological benefit for workers of having meaningful involvement in planning the firm's safety policy is that they begin to take ownership of safety outcomes, increasing their commitment to the program's successful implementation. This would also be expected to improve safety.

We measure this participation effect both through a numerical count of the types of decision-making activities the firm allows its workers to participate in (the extensive participation margin) and the degree of participation in those activities (the intensive participation margin). We expect greater involvement in either dimension will improve safety outcomes.

Another control variable measures the extent of information sharing that the firm engages in with workers concerning company finances, human resource planning, and workplace safety. Again, for reasons discussed in this section, we anticipate that as information sharing by the firm increases, safety will increase.

Lawler, Mohrman, and Ledford (1995) find in a survey of Fortune 1000 companies that 55 percent of the firms in 1987 and 48 percent of the firms in 1990 had improved workplace safety and health as a result of employee participation. Quantifying the magnitude of employee participation and the information employees contribute is not always easy, as other influences must be taken into account. Not only is the presence of one of these three dimensions of employee involvement (company finances, human resource planning, and workplace safety) likely to be correlated with the other two, but firms with employees involved in decision making are more likely to have employees involved in financial returns. Employee decision making may also be associated with the age of the workforce or the extent of workforce unionization. Since several correlated factors are simultaneously determining safety outcomes, to sort out the influence of any particular type of program requires multivariate controls for other types of programs and workforce demographics.

Regression analysis takes into account the influence of other factors included in the model. Definitive interpretation of prior research is hobbled in two ways: 1) one or more of the key HRM practices is omitted from examination (no prior study includes employee participation in both decision making and financial returns as well as measures of management safety culture), and 2) the variables are often examined in a univariate, rather than a multivariate, framework.

Past Research on HRM Practices

Shannon et al. (1996) is an example of a study limited in both of these ways. It doesn't have any variables on employees' involvement with the company's financial returns, and it doesn't present the results of the analysis in a multivariate framework. The researchers examine a matched sample of questionnaires sent to firms in Ontario with data on those firms' lost-time frequency rates (equivalent to the analysis of claim frequency given below). Survey questionnaires include responses from both workers and management. The paper discusses only univariate statistical analyses that compare various workplace practices to whether a firm has a low, medium, or high claim frequency. A multivariate regression analysis is mentioned but not reported on in the paper.

As discussed above, the absence of information about workers' involvement in financial returns makes it difficult to interpret the univariate correlations between safety outcomes and worker involvement in decision making. We must keep these cautions in mind when looking at studies like Shannon et al. (1996), which reports that lower claim frequency is associated with greater employee involvement in the firm's strategic decisions on safety. Specifically, claim frequency was lower where the level of worker participation was judged high, either by the workers or by the management. We took particular interest in the close agreement between workers' and managers' rankings of the relative importance of the various safety dimensions surveyed. Such close agreement in that survey suggests that our survey, which asked questions of senior human resource management, may have elicited much the same response if the questions had been asked of workers instead of managers.

Habeck, Hunt, and VanTol (1998), Habeck et al. (1998), Hunt and Habeck (1993), and Hunt et al. (1993)—extending the earlier research of Habeck (1993), Habeck, Leahy, and Hunt (1988), and Habeck et al. (1991)—relate the disability outcomes of 220 Michigan firms to those firms' HRM practices. This research is the first serious analysis of how management safety culture affects injury claims. In addition to collecting survey information for the 220 firms, Hunt et al. (1993) held extensive interviews with 32 of the 220 firms and find a qualitative difference in those firms that engage in what the researchers call a "participative culture." In other words, they find a qualitative difference between firms that facilitate employee involvement in decision making and those where the employees do not participate in the firm's decision making.

To capture the richness of the survey responses from the quantitative part of their investigation, Hunt and Habeck (1993) use factor analysis to create eight HRM policies and practices into which the responses fall. One of these factors, "people-oriented culture," is largely about employee involvement in firm decision making. Of the 12 variables loading into this factor, seven pertain to employee involvement or information sharing with employees:

1) "Working relationships are collaborative and cooperative in this company."

2) "Employees are formally included in the company's goal-setting and planning process."

3) "The company achieves open communications, where employees feel free to raise issues and concerns or to make suggestions."

4) "The company shares information with employees about the financial status and the productivity needs of the company."

5) "Management seeks and considers employee input in company decisions."

6) "Employee involvement programs, such as quality circles and labor-management participation teams, are used to generate employee participation in company operations."

7) "Workers have some control over work process and productivity demands."

This factor is not statistically significant in Hunt and Habeck's (1993) multivariate regression, but the results are suggestive: a 10 percent increase in this people-oriented culture variable is associated with a 4.2 percent reduction in lost workdays rate, as predicted by our model. This analysis focuses only on claim frequency, so the authors do not report any analysis of claim severity. Our results on claim frequency support the analyses of both Hunt and Habeck (1993) and Hunt et al. (1993), and suggest little or no effect of involvement in decision making on claim duration. Their findings with respect to management culture variables (the focus of their study) will be discussed below.

Eaton and Nocerino (2000) examine the impact of safety committees on injury claims in the New Jersey public sector. They find that the existence of a safety committee is associated with higher injury rates and speculate that this is probably because safety committees were initially established in workplaces that were the most dangerous. This sort of endogeneity may also explain the results of Fairris and Brenner (2001), who find that quality circles are associated with higher reported incidents of cumulative trauma injuries in an analysis of firms with three-digit SIC codes, whose firm-specific data on quality circles was matched with industry aggregates. (Teams and total quality management proved statistically insignificant in the analysis.)

While Eaton and Nocerino (2000) find the existence of safety committees associated with higher injury claim rates, they also find that greater worker involvement in safety committees leads to better safety outcomes. This is consistent with our expectations concerning employee decision making, including both the number of dimensions (decision making, financial returns, and safety process) that workers participate in and the degree of that participation.

Rooney (1992) finds that employee participation in decision making lowered the incidence of one or more Occupational Safety and Health Administration (OSHA) reportable injuries in his sample of 85 firms, although his analysis excludes firm characteristics and workers' average wages.

Grunberg, Moore, and Greenberg (1996) find that wood product firms with higher levels of worker decision making did no better than those with less worker decision making, contrary to the authors' expectations. One of the problems identified by the authors is that their sample of employee-owned firms also comprised the firms with the most precarious employment outlooks, so their measure of worker decision making may have been confounded with the effects of expected downsizing (which, for moral hazard reasons, may increase reported claims). Moreover, the authors' use of workers' self-reports on firms' safety may also have led to biased recall: conventional wood-product mill workers seemed to underreport their accidents, while cooperative workers seemed to overreport their accidents.

Rooney (1992) and Grunberg, Moore, and Greenberg (1996) compare employee-owned firms, in which employees are involved in both decision making and the firms' financial returns, with firms that are not employee-owned, in which neither type of participation takes place. The differential effects of decision-making participation relative to financial-returns participation are not separated. Unlike the Rooney (1992) and Grunberg, Moore, and Greenberg (1996) studies, Park (1997) estimates employees' participation effects by type, with one dummy variable indicating any involvement in the firm's financial return and another dummy variable indicating any involvement in the firm's decision making. The separation of these effects is theoretically desirable, although Park's empirical implementation is somewhat crude: these dummies distinguish neither the types nor the extent of participation within these broad groups. For example, a firm with only a suggestion system would receive the same value as a firm with total quality management, joint labor-management committees, and an employee representative on the board of directors.

Park merges Minnesota workers' compensation claims with survey data from the Minnesota Human Resource Management Practice file, developed at the Industrial Relations Center at the University of Minnesota. He finds that employee participation in decision making lowers the injury claim rate, as our model suggests, but that the reduction is not statistically significant.

Baril and Berthelette (2000) analyze correlates of early return to work for a sample of Quebec workers' compensation claims. Their qualitative analysis draws upon detailed interviews with 16 firms. Though no formal statistical analysis is drawn from these interviews, there appears to be a consensus among those interviewed that health and safety committees facilitate earlier returns to work, as does information sharing by management. However, the authors report that some types of worker involvement actually impeded returns to work: returns to work fell when rigid union seniority rules made it difficult to reassign workers to new jobs, and returns to work fell in firms where multiple unions were present. Multiple unions seem to impede temporary job reassignments for injured workers.

EMPLOYEE PARTICIPATION IN FINANCIAL RETURNS

Employees' participation in the financial returns of the company will change their safety incentives, but not unambiguously. Whether or not employees take appropriate precaution against injury risk depends in part on the incentives they have to do so, including the extent to which they will bear the cost of failing to take care. This is one reason that workers' compensation benefits do not fully replace lost wages. Since the claimant bears some of the lost wage costs, he has greater incentive to take care before an accident occurs and a greater incentive to return to work once an accident has taken place. Incentives are important because in many workplace accidents information is asymmetric: claimants' return-to-work capability is difficult to observe directly; often the injured worker knows more about his work ability than does the firm.

If employees take advantage of this informational asymmetry by changing their behavior because of insurance coverage, there is said to be moral hazard. For example, the extent of back pain is generally assessed through a worker's self-report of pain. If that worker's disability benefits were as high as his wages, and the worker didn't like either his job or his work supervisor, he might choose to stay away from the job longer after the onset of back pain symptoms than he would in the absence of insurance coverage. Partial insurance coverage, as is discussed above, is one way insurance attempts to limit such moral hazard, by making the insured employee bear some of the cost of being away from work.

Another mechanism for limiting worker moral hazard would be to involve the worker in the firm's financial returns. Employee participation in financial returns is measured by the number of programs through which the worker shares in the financial outcomes of the firm. To the extent that involvement with the firm's financial returns affects workers' expected income, they will tend to be more cautious if they can increase the firm's profits (and hence their income) by doing so. In experience-rated firms, for example, workers could reduce insurance premiums by filing fewer claims and could lower training costs and minimize reductions in output by missing fewer days of work.

Moral hazard may also be a problem with the firm: experience rating of the firm's insurance premiums (where a firm's future premiums depend on the current claims, so that higher-than-expected claims raise premiums) may induce firms to deny more claims than they would in the absence of experience rating, in order to reduce their insurance costs and increase their profitability. Sharing those profits with the employees through financial participation rights lowers the incentive for firms to engage in such moral hazard behavior as well.

Involving workers in the firm's financial returns also increases workers' willingness to provide information concerning effective changes in HRM policy and practice. To the extent that this increases the returns to safety investments, or lowers the costs of safety investments, accidents will fall. The demand for safety outcomes may also increase if participation in financial returns increases workers' wealth and thus lowers their willingness to bear workplace risk.

However, an increase in financial returns may increase accidents as well. If bearing more risk increases the expected profitability of the firm more than the perceived costs (i.e., working without a possibly cumbersome safety guard, or working long hours without rest), then the employee may actually have less incentive to take care, and injuries and injury claims may rise.

Finally, participation in financial returns may be ineffective in lowering workplace safety costs because of the free rider problem: if each worker perceives that his contribution to firm safety is negligible and if safety maintenance is costly, he will let others take care while he does not. But to the extent that others feel this way, no one takes care, and the effects of employee participation on safety outcomes will be muted. Employees will ride for free by benefiting from the system without contributing to it. For example, workers with lower back pain might not file a claim if they thought they would bear the full costs of the claim. But if they were to realize that they would gain the full benefits but share (indirectly) in only a fraction of the costs, they would file. So if the extent of ownership is too small to overcome the free rider problem, moral hazard claims may be filed even if there is some involvement with the financial returns of the company.

Employee participation in decision making may help to diminish the free rider potential by increasing workers' involvement and lowering the incentive for workers to engage in opportunistic behaviors: if employees' financial participation increases peer monitoring pressures, for example, free riding may be reduced.

Most research on employees' financial involvement with the firm centers on comparisons of employee-owned firms with firms in which the employee has no ownership interests; the research ignores other forms of employee involvement in the company's financial returns. Rooney (1992) and Grunberg, Moore, and Greenberg (1996), as reported in the last section, compare employee-owned firms that had employee participation in decision making with firms that lacked employee participation in financial returns or decision making. Rooney (1992) finds that employee ownership (along with employee participation in decision making, as discussed in the last section) lowered the incidence of one or more OSHA-reportable injuries among companies in his sample of 85 firms. Grunberg, Moore, and Greenberg (1996) find that wood product firms with employee ownership had no better safety outcomes than those without employee ownership. These researchers could not disentangle the effects of participation in financial returns from participation in decision making, and they only examined one dimension of participation in financial returns: whether or not the firm was employee-owned. In addition, these analyses don't control for any firm characteristics (though Grunberg, Moore, and Greenberg [1996] match firms in their sample) and may suffer from recall bias and sample selection bias, as noted in the previous section.

Unlike the Rooney (1992) and Grunberg, Moore, and Greenberg (1996) studies, Park (1997) distinguishes between decision-making participation and financial-returns participation in his study of Minnesota workers' compensation claims. Unexpectedly, Park found that employee participation in financial returns increased the injury rate, as did the interaction between financial returns and decision making. That is, as employee participation in the firm's financial returns rose, so did the injury rate, and the injury rate rose even more in firms with employee participation both in the firm's financial returns and in the firm's decision making. This is contrary to our theoretical expectations, as given above.[2]

In an analysis of 117 California firms, Hakala (1994) reports that workers' compensation experience modification fell as the degree of employee ownership increased. However, a firm's experience modification—an adjustment in its premiums according to whether the firm's losses are above or below the mean level in its risk class—depends on a firm's risk classification, making interclassification comparisons tenuous. Since the number of firm-level control variables is minimal in Hakala's research (as it is in Rooney's [1992] research), the findings need to be interpreted with caution.

MANAGEMENT SAFETY CULTURE

Management safety culture means the employer's commitment and leadership in making a safer workplace environment. In this research, our empirical measure of management safety culture is a Likert scale index of responses to the following issues:[3]

1) management's support for clear goals and objectives on safety and health policy,

2) management's leadership in setting goals on safety and health,

3) management's interest in safety and health issues as a part of the firm's strategic level of decision making,

4) management's willingness to share safety-related information with employees, and

5) management's commitment to reemployment of disabled workers and a return-to-work program for injured employees.

An increase in management safety culture should have the same impact on safety outcomes as an increase in worker participation in safety decision making, for similar reasons: as more management resources are employed toward integrating safety within overall corporate strategy—and as more ways are found to minimize post-injury return-to-work hurdles—accident costs will be reduced. To the extent this happens, the returns to safety investments increase, the level of job safety rises, and time away from work because of injuries falls. Higher values of our measure of management safety culture should be associated with greater safety outcomes.

Improved management safety culture has been shown to be associated with lower accident rates in some industries. Moses and Savage (1992, 1994) analyze the Federal Highway Administration's audit questions asked of all truck and bus companies. This audit consists of 57 yes-or-no questions that are used to rate carriers as satisfactory, unsatisfactory, or conditional. Unsatisfactory ratings trigger educational efforts and an additional, more detailed audit. Among the variables most predictive of accident rates are questions concerning the management safety culture of the carriers: Moses and Savage find that the lowest accident rates occur among carriers whose senior management report they are concerned with safety and communicate that concern to their employees.

Habeck et al. (1991) find that firms practicing poor disability management techniques had twice as many OSHA-recordable injuries but four times as many workers' compensation claims as firms practicing effective disability management. Hunt and Habeck (1993) reach a similar conclusion in their noted study of 220 Michigan companies. They find that management diligence concerning workplace safety, and policies providing for a proactive return to work, are associated with fewer disability claims and shorter disability durations. In particular, Hunt and Habeck's (1993) "Active Safety Leadership" factor focuses on many of the same elements as does our management safety culture; indeed, all 13 of the variables loading into this factor directly or indirectly deal with management safety culture as we have defined it.[4] The authors find that as Active Safety Leadership increases by 10 percent, claim frequency falls by 5.7 percent, a decline that is statistically significant. Our results on claim frequency support Habeck and Hunt's analysis, though our estimated response is larger than theirs. Moreover, we also find quantitatively large and statistically significant responses in claim severity, which reinforce the effect of lower claim frequency.

Hunt et al. (1993) also find that neither efforts to enhance wellness in workers nor ergonomic solutions prove to be effective in improving disability outcomes. In fact, ergonomic solutions to prevent injury and subsequent disability actually result in slightly higher disability rates. Although the result is odd, it is not statistically significant.

Shannon et al. (1996) report that lower claim frequency is associated with more management involvement in the firm's safety process. Specifically, claim frequency is lowest where management safety culture is strongest: where health and safety responsibilities are part of the manager's job description, where health and safety are an important part of his annual salary review, and where senior management attend the firm's health and safety committee meetings. That is to say, claim frequency is lower where management perceives itself—and is perceived by workers—as doing an effective job in supervision, and where management ranks health and safety as a high moral concern.

Claim frequency is also lower in the Shannon et al. (1996) sample, where a long-term disability plan is provided. The cost of such plans, whether by experience-rated premiums or by self-insurance, reflects the long-term safety experience of the workforce. Hence it is expected that where such programs are in place, management will be more attentive to workplace risks and will work harder to limit those risks. In other words, it will have an additional financial incentive to be involved with the safety culture of the firm.

Baril and Berthelette's (2000) interviews on early return to work for a sample of Quebec workers' compensation claims suggest that upper management support is important. When upper management support is lacking, safety committees are not able to exert any pressure on supervisors to encourage them to comply with the firm's health and safety rules.

Although a management culture of safety involvement may be established in places where there are perceived to be greater risk factors, the management culture in our research and in the research of Hunt et al. (1993) is assumed to have been determined by historical forces before the current period, and not by safety outcomes in the current period. We hope to minimize the possibility of reverse causality by including industry and occupational risk variables in the analysis when examining the impact of HRM programs. Following earlier research, we also treat the degree of employee participation and information sharing as exogenous in our analysis.

DOWNSIZING AND WORKERS' COMPENSATION

There are no studies using firm-level data examining how downsizing affects workers' compensation costs, though there have been several studies examining how individual claims are affected by employment status and how aggregate workers' compensation costs are affected by aggregate unemployment rates. Fenn (1981), using a British sample of claimants, reports that claim durations increase with higher levels of unemployment. Butler and Worrall (1985), relying on a U.S. sample of claimants, find that claim durations also increase from the incidence of layoff or unemployment. Fortin and Lanoie (1992) find that higher levels of unemployment are associated with higher frequencies of workers' compensation claims-filing as well, and it appears that workers undergoing jobless spells are migrating from obtaining coverage under unemployment insurance to obtaining coverage under workers' compensation insurance. Hartwig, Kahley, and Restrepo (1994) and Hartwig et al. (1997) find that the business cycle is a good predictor of workers' compensation costs. In particular, as the unemployment rate increases, workers' compensation claim frequency increases. Hence, layoffs and increased unemployment are likely to increase workers' compensation costs.

There are numerous reasons why downsizing might increase workers' compensation claims. Given the economic uncertainty associated with a reduction in the workforce, those who perceive that they may be laid off have some financial incentive to file a claim: while he is on a workers' compensation claim, a worker may still be eligible for other fringe benefits such as medical care for the rest of his family. Moreover, workers' compensation benefits replace two-thirds of a worker's lost wages tax-free, while unemployment insurance only replaces one-half of the worker's wages and may be taxed. In this situation, some of the increased costs may be the result of moral hazard (Butler and Worrall 1991).

There are reasons why even those who are not laid off may be affected by the reduction in force. Downsizing frequently increases job stress (Mishra and Spreitzer 1998), and stress is known to affect workers' health and their tolerance for pain. Just as some professional golfers can play with a sore back or some ballplayers can play with bad legs when they want to, workers who enjoy their work or who are otherwise well compensated for their work can work even if they are experiencing some level of discomfort. Once the rewarding, interesting work disappears or the work becomes stressful, then pain becomes more noticeable (Fordyce 1996; Smith and Carayon 1996). None of the effects listed so far represent changes in the real level of risk associated with work, so these effects are generally considered types of claims-reporting moral hazard. However, there may be real effects if the downsizing alters the demographic characteristics of workers (such as resulting in more younger or contingent workers), or it substantially alters working hours (in that, say, full-time employees are now asked to work much longer hours than formerly) (Capelli et al. 1997; Park 1997; Butler, Park, and Zaidman 1998). If younger and contingent workers are laid off, then injury claim rates will decrease after a downsizing (while claim duration may go up or down, depending upon the resulting composition of claims). If employees work longer hours and thus are subject to more fatigue, then job risk may increase and result in more injuries.

WHAT WE EXPECT TO FIND

This study goes beyond much of the earlier research and—following the approach of Hunt and Habeck (1993) and Hunt et al. (1993)—seeks to estimate the role of HRM practices in the determination of workers' compensation costs in a multivariate framework. It uses a workplace safety model that incorporates a wider variety of HRM practices than has been previously employed. In particular, it analyzes the impact of the three important dimensions of HRM practices on safety: employee participation in decision making, employee participation in financial returns, and the firm's management safety culture. In addition, this is the first study to consider the effect of each of these factors on claim frequency and claim severity, and to ask whether any observed change is the result of changes in technical efficiency or moral hazard (principal-agent) incentives.

As workers and managers get more involved with firm safety (and as HRM practices increase within a firm), we generally expect that workers' compensation costs will fall. Although participation in employee-owned firms yields inconclusive evidence, prior research on employee participation in decision making—aside from employee ownership—indicates that greater participation in decision making and more information sharing increases safety and lowers workers' compensation costs. We expect to find the same results in our data, not only for injury claim rates but for injury severity as well. However, our expectation about the positive impact of HRM practices applies to claims frequency results and to the overall change in expected costs per worker, but not unambiguously to changes in claim duration. The ambiguous effect on claim duration arises because a change in claim frequency may differentially alter the mix between long and short duration claims. The observed average duration may increase or decrease as a result.

The safety effect of employee participation in financial returns is theoretically ambiguous and—except for employee-owned firms—there is little prior research on participation in financial returns. The financial return estimates derived from employee ownership are fraught with problems of interpretation, as discussed above. We focus on variation in the degree of involvement in financial returns, aside from employee ownership. If firm profitability increases by taking more employment risks, then accidents and workers' compensation costs could rise. We don't expect, however, that that will happen; we expect that more financial participation by employees will lower workers' compensation costs through either of the channels outlined in Chapter 1: technical efficiency will increase as workers have an increased incentive to raise output, or moral hazard (principal-agent) outcomes will improve as workers have an increased incentive to lower costs.

Like employee participation in decision making, there is some evidence that improved management safety culture lowers the injury claim rate, and that greater executive involvement in the firm's safety processes lowers the rate of injuries. We expect the same to be true in our sample. Though this has not been examined in the earlier literature, we expect that as the management safety culture improves (as our measures increase), the severity of injuries will fall as well.

THE DATA

In July 1998, the Minnesota Department of Labor and Industry sent out a safety survey to 230 state firms that had applied for a program called the Minnesota Safety Grant Program. The Safety Grant Program awarded funds of up to $10,000 to qualifying employers for projects designed to reduce the risk of injury and illness to their workers through the purchase of new equipment and other physical improvements to enhance workplace safety. Other costs, including the cost of training, training materials, and labor costs generally, were not reimbursed under this program.

Qualifying employers under the Safety Grant Program must have been inspected by a qualified safety professional and have had a project consistent with the recommendations of the safety inspection. In essence, projects funded were expected to reduce the risk of injury or disease. Employers under this grant were required to be committed to the implementation of the safety project, including showing a willingness to match the grant money awarded. This requirement of a commitment may explain the relatively high reported scores on the management safety culture index: either the sample firms were from management teams deeply committed to on-the-job safety (not implausible, given their interest in applying for the grant), or they merely reported on the survey that they were committed, since the same state agency that supplied them with the grant money for their safety improvements also sponsored the safety survey.

Preference under the Safety Grant Program was given to firms with a significant employment presence in their geographical area, and to firms where jobs were at risk because of safety shortcomings. This tended to favor small and medium-sized firms established in less urban areas of the state.

Some 121 firms completed the survey forms, for a sample response rate of 53 percent. The survey consists of 11 sections, covering the general safety record of the company, company and employee characteristics, management culture, human resource practices, safety practices, and safety consultation programs provided by the Minnesota Occupational Safety and Health Administration (MNOSHA) consultation unit. Table 2.1 summarizes the question items included in the 22-page survey. These 121 firms have matched with about 5,125 workers' compensation indemnity claims for the years 1990 through 1998—that is, the claims were filed by workers of those 121 firms. Federal employer identification numbers were only available for about 10 percent of the sample and have proven to be unreliable for matching. Hence, we used the name of

Table 2.1 Contents of Minnesota Workplace Safety Practices (MWSP) Survey

Section	Section title	Content
Section I	Workplace Safety Statistical Data	Safety records and other general information of the respondent firm • SIC (Standard Industrial Classification) codes • OSHA 200 log information for the past five years • workers' compensation premium and covered payroll for the past five years • OSHA compliance, number of work sites
Section II	Company and Employee Characteristics	Information on employee characteristics and general employer characteristics • number of nonmanagerial workers • recent layoffs or cutbacks in employees • number of shifts operated and frequency of employee shift changes • workforce demography (age, education, tenure, gender, race), occupational mix
Section III	Management Culture and HRM Practices	Top management's commitment in safety and health policy and employee participation practices • management culture (management's strategic involvement in safety practices) • employee participation in safety practices (safety committees, union negotiations, ergonomics, performance review, certification requirements, return-to-work program, etc.) • employee involvement in firm decision making (quality circles, work teams, Total Quality Management, representation on the board of directors, etc.) • employee involvement in financial incentive programs (profit sharing, gain sharing, employee stock ownership plan, group bonus plan, 401(k), etc.)

Section IV	Safety Practices	Respondent firm's current safety practices and policies
Section V	Experience with Workplace Safety and Health Consultation Service	Respondent firm's experience with Safety and Health Consultation Service, including OSHA workplace consultation services
Section VI	Safety Committees	Respondent firm's experience with Labor-Management Safety Committee
Section VII	Minnesota Safety Grant Program	Respondent firm's experience with Minnesota Safety Grant Program
Section VIII	Minnesota Log Safe Program	Respondent firm's experience with Minnesota Log Safe Program
Section IX	Minnesota Workplace Violence Program	Respondent firm's experience with Minnesota Workplace Violence Program
Section X	Minnesota Safety and Health Achievement Recognition Program (MNSHARP)	Respondent firm's experience with Minnesota Safety and Health Achievement Recognition Program (MNSHARP)
Section XI	General Respondent Information	Marketing information about OSHA consultation programs

NOTE: In Section V, only general features of consultation services are covered. Individual-specific OSHA consultation programs are presented starting in Section VI of the MWSP Survey.

the firm in matching the two databases (i.e., matching the survey data with the workers' compensation administrative claims data).

The matching process involves three steps. First, we wrote an SAS program that screened the possible matches by searching over relevant character strings. If the firm were named "Park Inc.," for example, the SAS program would filter out all similar possible names, including "Parks Incorporated," "SPARK INC.," "PARK Inc.," "BALLPARK Inc.," etc. Second, a researcher at the Minnesota Department of Labor and Industry familiar with the Minnesota workers' compensation database checked each potential match. Third, we verified the matches by using the company's zip code from the addresses given in both data sets.

Following this procedure, we merge firm-level data from the survey to claimant-level data from Minnesota's workers' compensation files at the Department of Labor and Industry. Since costs are the product of claim frequency, claim duration, and benefits, we partition our statistical analysis into claim frequency and claim duration components to see whether the HRM practices affect claim frequency, claim duration, or both. This will provide evidence about whether costs are reduced either because of loss prevention effects (in that a particular practice reduces the number of claims) or loss control effects (in that a particular practice limits the costs of those injuries that have occurred). We assume that the benefit parameters (maximum and minimum benefits) are exogenous relative to the choices made by the firms in our survey and do not model benefit determination here.

As a result of our desire to partition HRM effects into their claim frequency and claim duration components, our descriptive sample statistics vary by type of analysis. These descriptive statistics are given in the empirical work in the next two chapters of the monograph, next to the corresponding analysis relevant to the descriptive statistics.

Notes

1. For notable exceptions, see Hunt et al. (1993), Park (1997), and Rooney (1992).
2. Park (1997) offers one possible explanation: "The firms with relatively more hazardous working conditions may implement various types of safety-enhancing efforts, including the practice of employee involvement plans in workplace safety issues" (p. 81)—in other words, reverse causality.
3. Likert scales were developed in 1932 by Rensis Likert, an organizational psychologist, as the familiar five-point bipolar response format that asks people to gauge their responses by how much they support or don't support (something), how much they are willing or not willing, etc., based on at least a five-response scale—often a seven-response scale is employed. At one end of the scale would be something like "Strongly support"; at the other end, "Strongly oppose." These scales are widely used in education, psychology, and sociology.
4. For more on the Active Safety Leadership factor, see also Habeck, Hunt, and VanTol (1998), Habeck et al. (1998), and Hunt et al. (1993).

3
Earlier Safety and HRM Practices

Employee Participation, Management Safety Culture, and Corporate Downsizing

CORPORATE DOWNSIZING AND HRM PRACTICES: THE SAFETY CONSEQUENCES

To gauge the influence of human resource management (HRM) practice on corporate safety, we estimate the direct and indirect effects of HRM practice on the firm's lost workday costs due to injury. In particular, we examine how lost workday costs vary along different HRM dimensions, as discussed in the last chapter. These dimensions are 1) the extent and intensity of employees' decision-making involvement and the level of employees' financial involvement in the company's performance, and 2) indices of management's commitment to workplace safety and the extent of management's information sharing. We calculate the estimated strength of these effects by their coefficients in claim frequency and claim duration regressions, then we relate the effects to their implied safety benefits.

The indirect effects of these programs may also be important. Anecdotal evidence suggests that injury claims tend to increase with announced reductions in workforce or as local unemployment rates rise. We estimate how layoffs affect claim duration by including a dummy variable indicating whether the firm has experienced any recent employment layoffs or cutbacks. In our models, we were also interested to see whether any of the corporate safety culture variables listed in the previous paragraph modify the layoff/claim-duration effect: that is, whether there are interactions between HRM practices and downsizing.

Competitive pressures to restructure the workplace by adopting new technologies, new management practices, new work tasks or processes, and new models of work organization give greater impetus to finding HRM practices that improve firms' performance (Appelbaum and Batt

1994; Cappelli et al. 1997; Levine and Tyson 1990). The search for improved performance not only has changed work processes for many employees, it also has led to changes in the size of the workforce itself. Corporations seek to become more cost-competitive by reducing their number of employees. Corporations also seek to become efficient by changing their corporate safety culture. For example, they may change the level of employee involvement or the types of safety programs they employ.

CLAIM FREQUENCY RESULTS

The descriptive statistics in Table 3.1 indicate a relatively high lost-work-time injury rate (.06) and a substantial amount of downsizing (32.6 percent had experienced downsizing of their workforce in the 12 months preceding the sample survey date), reflecting the dynamics of the midsize and small firms in the sample. Twenty-five companies are represented in Tables 3.1 and 3.2 in our analysis of claim frequency; some of these companies provide two or three years' worth of data. The relatively small sample size occurs because the "percentage of production workers" variable is missing in many of the surveys. We include percentage of production workers (and deal with the smaller sample size) in the analysis, however, because it is the best proxy we have for occupational differences (hence, intrinsic workplace risk) between firms, a factor we think is important to control for in our analysis, for reasons mentioned previously.

The values of the HRM variables indicate a relatively high rate of employee and management involvement in the safety efforts of the firm. The Management Safety Culture variable (MGTCULT) in Table 3.1 provides an index of management's relative involvement in the company's safety programs; the mean score of 23.01 indicates a relatively high level of management involvement in safety among these firms. Information Sharing (INFOSHR), another index, is created from six variables indicating the extent of information sharing with employees in several dimensions of firm activity: investments, production, human resource planning, profitability, corporate finance, and workplace safety. A score of 5 in each of these six dimensions indicates complete sharing; a score of 1 indicates no information sharing. The highest possible level of information sharing, then, would receive a score of 30; the lowest, a score of 5. The mean score of about 19 indicates that management shares a significant amount of information but not all of it.

Employee Participation in Decision Making (EPDM) is the sum of eight dummy variables indicating whether the firm allows employee participation in the following decision-making activities: suggestion system, quality circles, self-managing work teams, joint labor-management committees, quality of work life program, total quality management program, job redesign, and employee representation on the board of directors. The mean score of 2.63 indicates that the average firm in this part of the analysis uses less than three of these eight HR decision-making practices. Employee Participation in Decision Making (EPDM) indicates the breadth of decision-making activities that workers participate in, and Intensity of Employee Participation (EPDEG) indicates the depth. The mean score of 8.01 indicates that workers are reported to participate rather intensely in those decision-making activities they are involved with. So while workers are not involved in many strategic decision-making activities, they participate intensely in the few in which they are involved.

Similar to Employee Participation in Decision Making, Employee Participation in Financial Returns (EPFR) is the sum of 10 dummy variables indicating whether the firm allows employees to participate in the following: individual incentive plan, cash profit-sharing plan, gain sharing, pension plan, stock purchase plan, employee stock ownership plan, deferred profit-sharing plan, skill-based pay, 401(k) plan, and group bonus plan. The mean score of 2.125 indicates that, on average, the firms in the frequency analysis offer only two of these 10 options to participate in the financial returns of the firm.

Given the results cited in the literature above, our expectations are that higher values of these HRM variables will generally lower claim frequency. Which HRM practices actually do lower claim frequency, and by how much, is an empirical issue addressed in Table 3.2.

Since the results in the left-hand column of Table 3.2 come from a linear probability model, they are readily interpreted as the change in the probability of an injury given a unit change in the corresponding regressors. (Since the specifications in the center and right-hand columns involve injury severity measures rather than injury frequency, they will be discussed below.) For example, unionized workplaces have a 4 per-

Table 3.1 Descriptive Statistics: Firm-Level Analysis (1996–1998 Claims)

Variable	Definition	Mean	Std. dev.
Dependent variables			
INJINC (Injury incidence rate)	Number of workers' compensation claims per employee	0.060	0.256
INJSEV1 (Injury severity rate 1)	Number of lost workdays per employee	0.784	1.992
INJSEV2 (Injury severity rate 2)	Number of lost workdays per injury	18.616	31.425
Downsizing variable			
DOWNSIZE	A dummy variable coded 1 if the claimant's company experienced layoffs or cutbacks in the last 12 months, 0 otherwise	0.326	0.470
HRM variables			
EPDM	Number of employee participation programs in decision-making process	2.631	1.473
EPFR	Number of employee participation programs in the firm's financial returns	2.125	1.211
MGTCULT	Score of management's commitment to workplace safety	23.011	2.896
EPDEG	Score of the degree of employee participation in the company's decision-making process	8.011	1.256
INFOSHR	Score of the degree to which management shares information with employees on production issues	18.895	5.079

Control variables

UNION	A dummy variable coded 1 if the workplace is unionized, 0 otherwise	0.745	0.437
PCTWOMEN	Percentage of women employees	47.654	34.647
PCTAGE (25–54)	Percentage of employees aged 25–54	68.307	15.800
PCTPROD	Percentage of production employees	73.396	17.230

SOURCE: Authors' calculations.

Table 3.2 The Effects of Human Resource Management Policies on Workplace Injuries (Firm-Level Analysis: Full Model Adjusted for Asymptotic Covariances)

Variable	Dependent variable: Number of workers' compensation claims per employee	Dependent variable: Average lost workdays per employee	Dependent variable: Average lost workdays per injury
Constant	0.304** (0.141)	1.044 (2.883)	−50.065 (40.872)
Downsizing variable			
DOWNSIZE	−0.032** (0.019)	−0.072 (0.390)	3.753 (3.795)
HRM variables			
EPDM	−0.012** (0.006)	−0.119 (0.149)	−1.017 (2.428)
EPFR	−0.014*** (0.006)	−0.203** (0.115)	−2.154* (1.461)
MGTCULT	−0.012* (0.007)	−0.010 (0.165)	3.012* (2.052)
EPDEG	−0.002 (0.007)	−0.312** (0.163)	3.151* (2.111)
INFOSHR	0.005** (0.003)	−0.021 (0.097)	−0.813 (1.588)

41

Control variables

UNION	−0.040***	0.017	10.603*
	(0.017)	(0.459)	(6.785)
PCTWOMEN	−0.002***	−0.020	−0.031
	(0.0007)	(0.013)	(0.131)
PCTAGE (25–54)	0.0001	−0.029	−0.508*
	(0.0006)	(0.019)	(0.321)
PCTPROD	0.0009*	0.017	0.270**
	(0.0006)	(0.011)	(0.134)
Year dummies	yes	yes	yes
Industry dummies	yes	yes	yes
F value	2.608***	1.007	0.769
n	78	78	54

NOTE: Estimated heteroskedastic consistent standard errors are in parentheses. * significant at the 0.10 level (two-tailed test); ** significant at the 0.05 level (two-tailed test); *** significant at the 0.01 level (two-tailed test).
SOURCE: Authors' calculations.

centage point lower frequency of injuries than nonunion workplaces. A 10 percentage point increase in the number of production employees results in a 0.9 percentage point increase in the injury rate. An increase in the injury rate accompanying the addition of more production employees is expected, as production employees engage in riskier work activities than managerial workers. The union effect is more difficult to interpret: unions tend to enforce safety standards at work more than nonunion workplaces, which would explain why unions might decrease the injury rate. However, unions tend to form at worksites that are intrinsically more dangerous, so on the basis of sample selection one might expect a positive coefficient here.

The strongest effect among the control variables, however, is that of female workers. Nearly half of the workers in our frequency are female employees (47.65 percent). Our results suggest that increasing the number of females by 10 percentage points (to 57.65 percent) would decrease the injury rate by 2 percentage points. An increase of one full standard deviation in the percentage of female workers (that is, an increase of 34.6 percentage points) would decrease the injury rate by 7 percentage points.

The negative coefficient on the Downsizing dummy variable (DOWN-SIZE) suggests that sample selection and moral hazard mitigate the filing of additional claims. For those already on a lost-time claim, the effect of downsizing is to increase claim duration (as indicated by the 3.75 coefficient in the right-hand column of Table 3.2 for aggregated firm results and the positive Downsizing coefficients in Table 3.4 for individual claimant data). However, while claim duration increases for extant claims, claim frequency falls in those firms experiencing layoffs: either downsizing selectivity retains those workers least likely to file a claim (sample selection), or workers won't file claims for fear of being included in the next round of layoffs (i.e., there is a moral hazard that lowers the firm's safety costs). However, the downsizing effect is relatively small: downsized firms have only a 3.2 percentage point lower rate of injury than firms that have not downsized, though the difference appears to be significant at the 5 percent level.

Because HRM variables are our central focus in the analysis, we discuss those results from two different perspectives. The first is the standard regression interpretation of the estimated coefficients, reflecting the percentage change in injury rates given a unit increase in the value of the respective HRM practice. For example, a one-unit increase in Information Sharing increases claims by 0.5 percentage points, as shown in Table 3.2 in the left-hand column. However, because HRM variables are constructed as sums of dummy variables, one cannot always intuit what a unit increase means in terms of HRM practice. Therefore we also report the effects of the HRM variables from a high-HRM-use perspective, looking at the change from the usual values for HRM to the values of the respective HRM variables employed by the top 16 percent of firms (thus only about one-sixth of the firms have this level of HRM practice). This high HRM use is the level of HRM practice that lies one standard deviation above the mean. Hence, the claim frequency is 2.5 percentage points higher at the high-HRM-use level than at the average level, because a standard deviation increase in the value of information sharing from 20 to 25 percentage points increases claim frequency by 2.5 percentage points (5 × 0.005).

The human resource management practice variables in the left-hand column of Table 3.2 generally have the expected signs, including statistically significant impacts on the claim rate. The intensity of worker involvement (EPDEG) does not affect claim frequency (the − 0.002 coefficient is statistically insignificant).[1] More information sharing seems to increase the injury rate, as discussed in the last paragraph. Adding another program to involve employees in the firm's decision making (increasing EPDM by one unit) lowers the injury rate by 1.2 percentage points. Increasing the ways that employees share in the company's financial returns (EPFR) by one more program lowers the injury rate by 1.4 percentage points, and adding one more dimension to management safety culture (MGTCULT) also lowers the injury rate by 1.2 percentage points.

While Employee Participation in Decision Making, Employee Participation in Financial Returns, and Management Safety Culture all have roughly the same estimated coefficient, Management Safety Culture's influence is nearly twice as great when measured on a high-HRM-use perspective because variation in management safety culture is so much greater than variation in employee participation: in Table 3.1, the Management Safety Culture standard deviation is nearly twice as great as that of the employee participation variable EPDM and more than twice as great as that of EPFR. Hence, while the move to a high-HRM-use level would lower the injury rate by 1.7 percentage points for either of the employee participation variables, the move to a high-HRM-use level for Management Safety Culture would lower the injury claim rate by 3.5 percentage points. Finally, the move to a high-HRM-use level in the intensity of employee involvement in decision making (EPDEG) decreases the injury claims rate by 0.3 percentage points.

With the exception of Hunt et al. (1993), there are no studies of the impact of HR practices on claim frequency. Hunt et al. (1993) analyze survey results for 220 medium and large firms in Michigan and link their safety practices to their claim filing rates and measures of claim severity. Their safety diligence, safety training, proactive return-to-work program, and active safety leadership factors (which overlap our Management Safety Culture variable) are negatively correlated with their lost-workday case rate and with the workers' compensation claim rate in multivariate regressions and are generally statistically significant. Their sample sizes vary between 187 and 161, depending on specification. In the workers' compensation specification, Hunt et al. (1993) find that firm-size dummy variables are not usually statistically significant but that industry dummy variables are. They also report a regression, with aggregate, average lost workdays per case as the dependent variable, in which they find no statistically significant effects of safety practices on claim severity.

In the next section we examine the duration of individual claims for a larger number of claims than Hunt et al. (1993) had available, using individual claimant data from the Minnesota Department of Labor and Industry.

CLAIM DURATION RESULTS

Aggregate data: Table 3.2 revisited. While the most credible information on claim duration comes from the analysis of individual claim durations in Tables 3.3 and 3.4, the middle and right columns in Table 3.2 provide some alternative estimates on how HRM policies affect claim duration. The middle column of Table 3.2 corresponds most closely to the expected cost analysis given in Table 3.6, below, but is not always consistent with those results: increases in Employee Participation in Decision Making, Employee Participation in Financial Returns, and Management Safety Culture reduce expected losses, as they do in

Table 3.6, but the Management Safety Culture effect is relatively small and statistically insignificant, unlike the Management Safety Culture effect of Table 3.6. Moreover, increases in Intensity of Employee Participation or in Information Sharing lower expected costs in Table 3.2's middle column but have a positive effect in Table 3.6.

A specification that is most like the duration specification of this section is the right column of Table 3.2: they both model the duration of injuries, though the empirical results differ somewhat. Increases in Employee Participation in Financial Returns and in Information Sharing are found to be linked to lower claim durations in Table 3.2, as they are in the results reported in Table 3.4 below. Intensity of Employee Participation has the same positive impact on duration in Table 3.2 as it does in Table 3.4. However, Management Safety Culture and Employee Participation in Decision Making have the opposite signs, though the Employee Participation in Decision Making variable is statistically insignificant in both specifications.

The estimated effect of Management Safety Culture on claim duration is 10 times larger in Table 3.4 than it is in Table 3.2, perhaps because those tables represent a slightly different sample of firms, or perhaps because the measure of claim duration is slightly different. The sample in Table 3.2, as previously noted in the discussion of the descriptive statistics, is limited to those reporting a full set of control variables (union or nonunion, percentage women, percentage young, percentage production employees). Thus there is a smaller sample of firms contributing to the estimates in Table 3.2 than in Table 3.4. However, we doubt, given the approximate similarity in their sample means, that this explains the differential response.

A more likely explanation of the difference in results is that claim duration is measured differently. The firm-specific OSHA data (reported to the U.S. Department of Labor) in the far right-hand column of Table 3.2 include workers whose injury duration is shorter than three days, whereas the data in Table 3.4 exclude those who don't satisfy the three-day waiting period. In workers' compensation, claims with a duration of less than the waiting period are called medical-only claims. Appel and Borba (1988) report that the medical-only category accounts for 81 percent of all claims, although it accounts for less than 6 percent of total costs. If Management Safety Culture successfully reduced long-duration claims while encouraging the reporting of short-term injury in-

cidents (which included medical-only workers' compensation claims), then this would explain both results: few long-duration claims occur as Management Safety Culture increases (Table 3.4), but more diligent reporting and monitoring of minor injuries does occur (Table 3.2). Unfortunately, the Minnesota data do not record information on medical-only claims, so pursuit of this hypothesis (claims shifting as Management Safety Culture changes) will have to wait for future research.

Individual claim data: Tables 3.3 and 3.4. The descriptive statistics for the duration analysis given in Table 3.3 indicate that the mean duration of workers' compensation claims among claimants with some lost-time workdays is 55.78 days.[2] However, 20 percent of these are right-censored (and were still in progress when we drew our sample of claims), hence the 55.78 days is an underestimate of the average completed spell of workers' compensation. Some claims, often those of longer duration, will still be in progress when information about the claims is gathered. If a claim has been in progress for two weeks, for example, when the sample of claims is drawn, we cannot be sure whether the final duration of the in-progress claim will be 15 days, 15 weeks, or even 15 months. We only know that the duration is at least two weeks long. If the duration of this in-progress spell is recorded as "2 weeks," we underestimate the final duration and bias the results. If the in-progress spells are thrown out, we tend to bias the estimates again, as these claims will tend to be of longer duration. Maximum likelihood estimation allows us to include these in-progress spells while accounting for the right-censoring in a way that doesn't lead to biased estimates. The maximum likelihood estimates, employing a Weibull duration model (McDonald and Butler 1990), are reported in Tables 3.4 and 3.5.

The values of the variables for employee participation and management culture in Table 3.3 indicate a relatively high rate of employee and management involvement in the safety efforts of the firm and are very similar to those reported in Table 3.1. The differences between the statistics in Table 3.1 and those in Table 3.3 stem from the weights given to different firms: in Table 3.1, each firm receives equal weight in determining the mean; in Table 3.3, the larger firms have more claims and so implicitly get greater weight than they do in Table 3.1. The main difference in the Employee Participation in Decision Making, Employee Participation in Financial Returns, and Information Sharing variable averages—all of which are more representative of the larger firms—is that HRM variables in the claim duration specification have slightly higher values in Table 3.3 than these same variables have in the claim frequency sample in Table 3.1. Given the results cited in the literature above, our expectations are that higher values of these HRM culture variables will lower claim duration. Which ones, and by how much, is an empirical issue addressed in Tables 3.4 and 3.5, where we employ maximum likelihood to estimate the duration of workers' compensation claims.

We generally find statistically significant effects, of the expected sign, in the duration models. While we do not report the estimated shape parameters of the Weibull distribution in Tables 3.4 or 3.5, they indicate negative duration dependence: as claim duration increases, the rate of exit from claimant status falls. Hence, the longer a claimant stays on a workers' compensation claim, the less likely he is to leave it.

The demographic variables in the Weibull regressions of Tables 3.4 and 3.5 have their expected effect. For example, a 10 percent increase in the age of a claimant increases the expected claim duration by 7 percent when there are no controls for industry, occupation, and injury, and it increases the expected claim duration by 4 percent when such controls are present (see the right-hand column coefficient for age).[3] This is the same sign, but twice the magnitude, of the age elasticity reported by Butler and Worrall (1985). The replacement rate elasticity is also higher here than that reported in Butler and Worrall (1985) but is within the range generally found in empirical research (Butler, Gardner, and Gardner 1997). In our Minnesota sample, a 10 percent increase in benefits (holding wages constant) increases the expected claim duration by about 5 percent (the right-hand column in Table 3.4). We find no gender differences in duration, nor do we find a self-insurance effect once industry and occupation are held constant.

The empirical findings in Table 3.4 also show Downsizing and HRM effects of the expected sign. If the firm has downsized its workforce in the last 12 months, then claim duration is 20 percent higher than it would be without any recent downsizing. These results are consistent with earlier findings that workers' compensation costs increase in environments where there is greater employment uncertainty. This finding is important since it is the first time that firm-specific downsizing information (as opposed to individual employment status or local unemployment in-

Table 3.3 Duration Model Using Claimant Data: Descriptive Statistics

Variable	Definition	Mean	Std. dev.
Dependent variables			
NWSPELL	Duration of nonwork spell: days of temporary total disability benefits paid in workers' compensation system[a]	55.78	140.91
DENIAL	A dummy variable coded 1 if claim was denied for liability by insurer, 0 otherwise	0.20	0.40
Downsizing variable			
DOWNSIZE	A dummy variable coded 1 if the claimant's company experienced layoffs or cutbacks in the last 12 months, 0 otherwise	0.40	0.49
HRM variables			
EPDM	Number of employee participation programs in decision-making process	3.12	1.36
EPFR	Number of employee participation programs in the firm's financial returns	3.08	2.16
MGTCULT	Score of management's commitment to workplace safety	22.62	2.29
EPDEG	Score of the degree of employee participation in company's decision-making process	8.12	1.00
INFOSHR	Score of the degree to which management shares information with employees on production issues	20.26	5.24

Claimant characteristics

MALE	A dummy variable coded 1 if claimant is male, 0 if female	0.53	0.50
LnAGE	Log of age of claimant at time of injury	3.63 $(39.47)^b$	0.31 $(11.54)^b$
LnRRATE	Log of wage replacement rate in Minnesota workers' compensation system[c]	4.28 $(73.40)^b$	0.19 $(14.41)^b$
SELF	A dummy variable coded 1 if claimant's company is self-insured for its workers' compensation coverage	0.29	0.45

[a] The dependent variable NWSPELL is transformed into logarithmic form in the LIFEREG function of SAS.

[b] Descriptive statistics of variables without taking logs.

[c] Real-wage replacement rate was used to capture both wage and expected workers' compensation benefit effects on the dependent variable. In accordance with the Minnesota workers' compensation law, RATE was calculated by the following formula (Minnesota WC income benefit schedule used; 1992 analysis of workers' compensation laws, U.S. Chamber of Commerce):

$$\text{RATE} = \begin{cases} \text{MAX}_t / \text{Wage} & \text{if } (\text{Wage} \times 0.66) \geq \text{MAX}_t \\ 0.66 & \text{if } [\text{MIN}_t \leq (\text{Wage} \times 0.66) < \text{MAX}_t] \\ \text{MIN}_t / \text{Wage} & \text{if } [(\text{MIN}_t \times 0.66) \leq (\text{Wage} \times 0.66) < \text{MIN}_t] \\ 1 \text{ otherwise,} \end{cases}$$

where Wage is average production employee's gross weekly wage, MAX_t is maximum amount of wage replacement through Minnesota workers' compensation system, and MIN_t is minimum amount of wage replacement through Minnesota workers' compensation system. (In the analysis of this study, log of RATE × 100 was included in the models as RRATE.)

SOURCE: Authors' calculations.

Table 3.4 Weibull Estimates of the Duration of Nonwork Spells in the Workers' Compensation System: The Direct Effects Model (standard error in parentheses)

Variable	Dependent variable: log of nonwork spell in Minnesota workers' compensation system	
	Model (1)	Model (2)
Constant	2.96**	4.02***
	(1.25)	(1.46)
Downsizing variable		
DOWNSIZE	0.25***	0.21**
	(0.09)	(0.10)
HRM variables		
EPDM	0.05	0.06
	(0.03)	(0.04)
EPFR	−0.13***	−0.13***
	(0.04)	(0.05)
MGTCULT	−0.14***	−0.15***
	(0.03)	(0.04)
EPDEG	0.20***	0.19***
	(0.06)	(0.07)
INFOSHR	−0.05***	−0.06***
	(0.01)	(0.01)
Claimant characteristics		
MALE	0.03	−0.13
	(0.08)	(0.09)
LnAGE	0.69***	0.41***
	(0.12)	(0.13)
LnRRATE	0.34	0.53*
	(0.23)	(0.28)
SELF	−0.41***	0.04
	(0.09)	(0.12)
Control variables		
Industry dummies	no	yes
Injury type dummies	no	yes
Occupation dummies	no	yes
Year dummies	no	yes

	Model (1)	Model (2)
Log likelihood for Weibull	−3057.40	−2655.01
n	1906	1732

NOTE: * significant at the 0.10 level (two-tailed test); ** significant at the 0.05 level (two-tailed test); *** significant at the 0.01 level (two-tailed test).
SOURCE: Authors' calculations.

Table 3.5 Weibull Estimates of the Duration of Nonwork Spells in the Workers' Compensation System: The Indirect Effects Model (standard error in parentheses)

Variable	Dependent variable: log of nonwork spell in the Minnesota workers' compensation system	
	Model (1)	Model (2)
Constant	3.41***	5.13***
	(1.31)	(1.55)
Downsizing variable		
DOWNSIZE	−.77	−.19
	(1.68)	(1.80)
HRM variables		
EPDM	.03	.005
	(.04)	(.05)
EPFR	−.04	.06
	(.06)	(.07)
MGTCULT	−.17***	−.20***
	(.05)	(.05)
EPDEG	.19**	.15
	(.08)	(.09)
INFOSHR	−.05***	−.04***
	(.01)	(.01)
HRM × Downsizing interactions		
EPDM × DOWNSIZE	.07	.14*
	(.07)	(.08)
EPFR × DOWNSIZE	−.13	−.28***
	(.08)	(.10)
MGTCULT × DOWNSIZE	.04	.02
	(.08)	(.08)
EPDEG × DOWNSIZE	.01	.08
	(.11)	(.13)
INFOSHR × DOWNSIZE	.003	−.02
	(.02)	(.02)
SELF × DOWNSIZE	−.34	−.25
	(.21)	(.24)

Table 3.5 (continued)

Variable	Dependent variable: log of nonwork spell in the Minnesota workers' compensation system	
	Model (1)	Model (2)
Claimant characteristics		
MALE	−.01	−.15
	(.08)	(.09)
LnAGE	.67***	.41***
	(.12)	(.13)
LnRRATE	.38	.43
	(.23)	(.28)
SELF	.028	.12
	(.04)	(.13)
Control variables		
Industry dummies	no	yes
Injury type dummies	no	yes
Occupation dummies	no	yes
Year dummies	no	yes
Log likelihood for Weibull	−3,054.20	−2,646.97
n	1,906	1,732

NOTE: * significant at the 0.10 level (two-tailed test); ** significant at the 0.05 level (two-tailed test); *** significant at the 0.01 level (two-tailed test).
SOURCE: Authors' calculations.

surance rates) has been used in an analysis of individual claim duration. The Downsizing effect is not only large in magnitude but statistically significant as well. Our results suggest that the tendency for workers' compensation costs to increase during downsizing operates through a claim duration effect rather than through a claims filing effect.

All of the HRM variables except Employee Participation in Decision Making (EPDM) also significantly reduce claim duration, and they generally have larger estimated effects on claim duration than they do on claim frequency. Even though the number of decision-making activities does not affect claim duration, the intensity of participation does. A standard deviation increase in the intensity of employees' involvement, creating a jump to the high-HRM-use level, increases claim duration by 20 percent. (Recall that Intensity of Employee Participation is the sum of two Likert scores indicating to what degree workers feel as though they "always" have control over their job tasks and whether they "always" participate in employee involvement programs.) The greater the level of intensity of employee involvement, the higher the safety costs. Our guess is that as workers become more involved in making and disseminating company safety policy, they are able to reduce the number of shorter, less serious injury claims more than they reduce the number of longer claims. This changes the mix of claims observed, increasing the observed duration of the remaining claims. If this were the only factor operating, however, the frequency effects would outweigh the duration effects, and overall costs per employee would fall as information sharing increased. We shall see that this is not the case. Other factors must be involved—future research will likely shed more light on this interesting result. One possibility is that as more information is shared in the company, workers are more willing to provide information on their workplace injuries, including minor injuries not involving significant loss of work time claims. Another possibility, which we find less plausible, is endogeneity bias: firms are more willing to share information in situations where risk is greatest.

We again report the effects of the HRM variables from the high-HRM-use perspective, looking at the change from the average HRM practice to the top 16 percent of high-HRM-use firms. From this perspective, if the number of financial returns programs (such as profit sharing) increases by a standard deviation by going from three such programs (the average) to five (the high-use level in our sample), then claim duration falls by 26 percent. High-use implementation in Information Sharing, which for our sample means going from an index score of 20 to 25, decreases claim duration by 30 percent. The greater the workers' financial interest in their firm, the lower their claim durations will be, on average. This reinforces the reducing effect that participation in financial returns has on claim frequency.

Finally, it appears that those managers most involved in their company's safety efforts achieve the greatest decline in the duration of their employees' claims. When an average firm employs high-use HRM practices—equivalent to a standard deviation increase in the Management Safety Culture index, from about 23 to 25—claim duration decreases by 30 percent. This is a large effect, and it reinforces the management safety effect on claim frequency. This is also consistent with the findings of Hunt et al. (1993).

These large magnitudes indicate a substantial direct impact of HRM practices on claim duration. Since the research cited above, as well as anecdotal evidence from the insurance industry, indicates that downsizing and employment cycles have a significant impact on claims filing, we also examine, in Table 3.5, whether there are indirect effects of HRM practices that operate through a reduction in the Downsizing variable. We measure this indirect effect by adding interactions between downsizing and HRM practices to the analysis.

In general, there is little evidence of such indirect effects. In Table 3.5, only the financial returns interaction in the right-hand column is statistically significant at the 5 percent level; all of the other interactions are insignificant at the 5 percent level. Taken as a group, the interactions in the left-hand column of Table 3.5 are statistically insignificant when using a likelihood ratio test, and the ones in the right-hand column are significant at the 5 percent level but not at the 1 percent level.

Hence, with the single exception of the Financial Returns × Downsizing interaction, the effects of HRM practices on claim duration are direct effects and do not seem to ameliorate the impact of downsizing on costs. The estimated negative effect of Financial Returns × Downsizing on claim duration indicates that when employees participate in the financial success of a company, they tend to file claims of shorter duration in the face of downsizing efforts, as would be expected. However, better HRM practices in other dimensions (Information Sharing, Management Safety Culture, and Employee Participation in Decision

Making) do not seem to mitigate the effect that Downsizing has on an increase in claim duration. It may well be that those who perceive that they are going to be downsized are no longer concerned about their participation in HRM practices, or about gaining information about those practices, because they see them as largely irrelevant. If they are no longer employed in that firm, firm outcome matters less to them.

COST IMPLICATIONS OF THE FREQUENCY AND DURATION ESTIMATES

Since expected indemnity costs per worker are simply the product of claim frequency, of claim duration (in weeks), and of the legislated benefits, we can examine whether safety programs operate through a loss prevention (frequency) effect or a loss reduction (duration) effect. Our results are summarized in Table 3.6. In our sample, the average indemnity cost (the amount paid to injured employees for lost wages) per injury was $175.50.[4] This average annual indemnity cost per employee can be calculated by taking the product of the number of claims per employee times the average duration (in weeks) per claim times the average statutory weekly benefit. To calculate the impact of a particular HRM practice variable (say, the effect of Employee Participation in Financial Returns, or EPFR) on annual indemnity cost per employee, we take the derivative of costs (in natural logarithm terms) with respect to that variable:

$$(3.1) \quad \frac{\partial \ln(\text{Cost})}{\partial \text{ EPFR}} = \frac{\partial \ln(\text{Frequency})}{\partial \text{ EPFR}} + \frac{\partial \ln(\text{Duration})}{\partial \text{ EPFR}} + \frac{\partial \ln(\text{Benefit})}{\partial \text{ EPFR}}$$

On the right-hand side of the above equation, the first term represents the percentage change in injury frequency relative to a unit change in the Employee Participation in Financial Returns index; it equals the estimated linear probability coefficient in Table 3.2 divided by the average probability of a claim. This is given in the left-hand column of Table 3.6. The second term on the right-hand side of Equation (3.1) represents the percentage change in injury duration relative to a unit change in the Employee Participation in Financial Returns index. It equals the estimated regression coefficient from the survival analysis for the duration

Table 3.6 Safety Cost Implications of Lost-Time Pay in Workers' Compensation When Adopting Various Human Resource Practices

HR practice	Impact on claim frequency (%)	Impact on claim duration (%)	Overall impact on indemnity costs (%)	Indemnity cost change per employee ($) (per std. dev. change)
EPDM	−20	6	−14	−24.57 (−36.19)
EPFR	−23	−13	−36	−63.18 (−76.51)
MGTCULT	−20	−15	−35	−61.43 (−177.89)
EPDEG	−3	19	16	28.08 (35.27)
INFOSHR	8	−6	2	3.51 (17.83)

NOTE: Based on 1996 average lost-time workers' compensation costs per employee of $175.50 in our sample (12 percent of claims were still open).
SOURCE: Authors' calculations.

model results given in Table 3.4. These coefficients are reproduced in the second column in Table 3.6. The third term on the right-hand side of Equation (3.1) captures the influence of a firm's Employee Participation in Financial Returns index on statewide statutory benefits. This effect is probably close to zero since there is no reason a change in employee involvement in the firm's financial returns—or any other HRM practice—should change the legislated benefit level.[5]

The overall change in indemnity costs can be approximated by multiplying the percentage change in the indemnity costs shown in the two left-hand columns of Table 3.6 (and summed in the third column) by the average indemnity cost in this sample, $175.50. This calculation is made in the far right-hand column of Table 3.6, both for a one-unit change in each of the HRM indices and also (in parentheses) for a one-standard-deviation change in HRM indices, indicating a movement to the high-HRM-use practice level.

As an example, if a firm were to go from using two to using three financial returns programs (Employee Participation in Financial Returns went from 2 to 3), then claim frequency would fall by 23 percent and claim duration would fall by 13 percent. Lost work-time benefit costs would fall by 36 percent (the combined impact of frequency and duration outcomes), which when multiplied by the average indemnity cost per employee of $175.50 yields a cost reduction in lost-time expenses of $63.18 per employee per year. This is the approximate reduction in per-employee costs that is achieved by adding another plan in which the employee shares in the financial returns of the company. If the employer were to add 1.211 more programs (that is, add one standard deviation more to Employee Participation in Financial Returns programs—Table 3.1) to the firm's current offerings, then costs would fall by 1.211 × $63.18, or by $76.51. A standard deviation increase is equivalent to moving from the average HRM offering (an HRM offering as good as about half of the firms) to an HRM offering that is in the top 16 percent of the firms.

Management Safety Culture is the HRM practice with the largest cost savings per employee. A firm that increases its level of involvement so that the Management Safety Culture variable goes up by 3—or increases by more than 10 percent and comes up into the high-HRM-use level—saves about $180 per year per employee. In a firm with 100 employees, this is an annual savings of $18,000 in lost-time pay alone. This is likely a lower bound on the benefits derived from improving management safety culture. Medical costs will be saved in addition to the lost-time pay costs; lost workdays will be eliminated and specific human capital thereby retained; and employees' level of job satisfaction may increase as well. There is a lot of variation in the value of this variable in our sample, as one can see by looking at the large standard deviation values in Tables 3.1 and 3.3. As there is a lot of variance in Management Safety Culture, there is a lot of opportunity for firms to lower safety costs by becoming more involved with work safety.[6] If, in a 100-employee firm, it costs less than $18,000 to become more concerned with safety processes and outcomes, then it is clearly advantageous for the firm to do so.

It is worthwhile to reemphasize that our estimate of the reduction in workers' compensation costs from engaging in these HRM practices is probably a lower bound estimate of the potential benefits. To the extent that the workplace is safer, either because physical risks have been reduced or because workers are taking more appropriate safety precautions, then some other accident costs are likely to be reduced as well. Uncompensated wage loss and pain and suffering associated with injuries will decrease, reducing the compensating wages paid to work-ers. Perhaps equally important in our highly skilled labor market, more firm-specific human capital can be retained as the number of days out due to injury falls.

Notes

1. EPDEG does appear to be associated with greater claim severity (the significant EPDEG coefficients in the center and right columns). An increase in EPDEG also is estimated to increase claim duration in the individual claimant analysis in Table 3.4, as we will show shortly.

   The models in Table 3.2 were estimated using three-year averages (the most current three years of data) rather than using cross-section/time-series analysis, as was employed in Table 3.2. The resulting coefficients were nearly quantita-tively identical to those reported in Table 3.2.

   We also estimated the models in Table 3.2 by including firm size as a regres-sor. It was statistically significant and negative in the left and center specifications of Table 3.2; its inclusion did not change any of the signs of the other regressors but did increase the magnitude of the MGTCULT, EPDEG, and DOWNSIZE coefficients, while slightly reducing the coefficients for EPDM, EPFR, and INFOSHR. We exclude it from our preferred estimation because of concern for potential endogeneity in the model (since firm size is also in the denominator of the dependent variable).

   Finally, interactions in the claim frequency reported below between the DOWNSIZE dummy variable and HRM practices were statistically insignificant (partially, of course, because of the small sample sizes).

2. The statistics here pertain to the larger claims denial regression sample. Howev-er, the smaller duration sample has mean statistics (for the independent variables) very close to those reported in Table 3.1.

3. There were seven occupational-dummy variables, seven year-dummy variables, and eight industry-dummy variables included in the model. There were also dummy variables for the following injury types: back sprain and strain, other sprains and strains, fractures, contusions and concussions, and lacerations.

4. $175.50 was the average in 1996 for our sample. 1998 and 1997 data were judged to have too many open claims to be used as a measure of costs per worker: 1998 had 52 percent of claims still open, while 1997 had 45 percent of claims still open. For 1996, only 12.6 percent of claims were open. We didn't use earlier years (with still fewer open claims) because of a concern that they might be less relevant for the survey results and because of cost-of-living differences.

5. The expected weekly benefit depends on the legislated maximum and minimum benefits, the replacement rate (these three benefit parameters are determined by the individual state legislatures), and the worker's wage rate. Since maximum benefit payments are low enough that most workers are constrained by the maxi-mum weekly benefit, a change in any HR practice won't affect the benefit they

receive—they will get the legislated maximum payment. However, workers below the maximum may have their benefits indirectly affected by the HR practices that are adopted if they value those practices. Workers who value the work environment generated by some of these HR practices, and whose weekly benefit would be below the maximum legislated benefit level, will take an implicit wage reduction because of the valued HR practices. Furthermore, because of the wage reduction, they will have their benefits indirectly affected by the presence of the HR practice. The size of the implied wage reductions necessary to fund the HR practices considered would not be enough to substantially alter the conclusions in Table 3.6. The greatest wage reductions, for example, would be for employer's share of profit sharing (or, perhaps, health and pension contributions). If this leads to a 3 percent reduction in wages, expected benefits would fall (for those between the minimum and maximum benefit levels) by 2 percent. Such a change is dwarfed by the changes given in Table 3.4. Finally, even if these changes are more substantial than we anticipate, note that they work to reinforce the conclusions drawn concerning the outcomes in Table 3.4, as they tend to further reduce workers' compensation costs.

6. Hunt et al. (1993) reach similar conclusions. In their survey, active safety leadership, safety diligence, proactive RTW program, and safety training proxy some of the same dimensions that we have included under Management Safety Culture. In their study, they also find that these are effective mechanisms through which to promote health and safety, and their empirical data suggest that there is also a relative abundance of variation in these practices across their sample of relatively large firms (Hunt et al. 1993, Table 4.2).

4

Reduced Moral Hazard
or Increased Efficiency?

Evidence from Claim Types and Claim Denials

The distinction made between the direct and indirect effects of HRM practices in the last chapter is not only intrinsically interesting, it suggests that HRM practices do not simply reduce claims-reporting risk (a change in the reporting of claims, even where real risk is held constant) but rather reduce real risk (through lowering the degree of intrinsic physical risk of employee risk-taking behavior). In Chapter 1 we argued that HRM practices may change real risk by more fully involving employees in the firm's strategic decision making or financial returns. Either of these measures increases employees' incentives to design and implement safety policies that improve workplace efficiency or at least help reduce impediments to the flow of information about safety risks between the firm and the employees. We explore this intriguing possibility further in this chapter by examining how HRM policies affect claim denials and the distribution of injury risks.

CLAIMS-REPORTING AND RISK-BEARING MORAL HAZARD

In the last chapter, we found that HRM practices reduce claim duration and, generally, claim frequency. However, this indicates nothing about whether the effect represents a reduction in real risk taking or simply a reduction in the propensity to report a claim. The former is called *risk-bearing moral hazard* while the latter is known as *claims-reporting moral hazard* (Butler and Worrall 1991). Though either kind of moral hazard increases an employer's costs, the distinction between risk-bearing and claims-reporting moral hazard has important implications for resource allocation. If changes in claim frequency and claim duration stem purely from the effects of reporting, without any change in the degree of intrinsic physical risk, then we would not expect to see changes in compensating wages for unreimbursed wage losses.[1] Further, if it is a pure reporting effect, then changes in safety programs or ergonomic standards are likely to have little impact on workers' compensation costs. If increases in workers' compensation costs are driven by reporting effects, additional expenditures may be more productive as subsidies to pension benefits or wellness programs than as increases in safety training.

The estimated direct effects in the previous chapter only indicate that downsizing increases workers' compensation claim duration and that most HRM practices decrease workers' compensation costs. These estimates do not indicate whether the corresponding changes in workers' compensation outcomes are a result of changes in the propensity to report claims or a result of changes in the degree of risk-taking behavior of workers. The propensity to report claims would be affected if participation in strategic decision making or in the firm's financial returns internalized safety costs to the employee so that he made the same decisions he would make if he were spending his own money. If claims-reporting behavior were important, then as employee participation in decision making and financial returns increased, the impact of downsizing on claim frequency or claim duration would fall. In other words, we would expect negative and statistically significant interactions between Downsizing and various HRM practices. This suggests that we may find some evidence on the reporting versus real risk explanations of cost changes by examining indirect evidence—that is, evidence besides the main HRM and Downsizing effects.

Some indirect evidence was provided in Table 3.5 of the last chapter, though it was not discussed as a test for claims-reporting moral hazard. To test the claims-reporting versus risk-bearing explanations using Table 3.5 coefficient estimates, we have to assume either that downsizing increases workers' compensation claim duration because it is a reporting effect (this is the traditional explanation in the empirical literature) or that downsizing represents a change in real risk that is due to job stress or job fatigue as the work pace increases. We now explore both of these interpretations of Table 3.5.

Downsizing outcomes as a claims-reporting effect. In the midst of downsizing, there are a number of reasons to suspect claims-reporting moral hazard, particularly if the claimant believes that he may be one of those losing a job. Workers' compensation claim status is preferred to unemployment insurance claim status for a number of reasons: workers' compensation benefits are higher than unemployment insurance benefits; workers' compensation benefits are tax-free and unemployment insurance benefits are not; and workers' other fringe benefits continue while they receive workers' compensation benefits, whereas there are few, if any, fringe benefits for unemployed workers. If the Downsizing effect is a reporting effect, HRM practices that increase employee involvement should mitigate these responses.

Downsizing outcomes as a risk-bearing effect. In this case, employment volatility and longer work hours associated with downsizing would either increase the intrinsic workplace risk or induce workers to change their risk-bearing behaviors. Downsizing would increase claim frequency, but this impact would be mitigated in those firms with better work safety conditions.[2] Since higher values of the Management Safety Culture variable (management's commitment to workplace safety) indicate better safety environments, then if risk-bearing moral hazard is reflected in the Downsizing effect, there should be a negative interaction between Management Safety Culture and Downsizing. Safer workplaces will reduce the risk-bearing effects associated with Downsizing.

The evidence. The Management Safety Culture × Downsizing interaction in Table 3.5 of Chapter 3 (MGTCULT × DOWNSIZE), instead of being statistically significant and negative, is insignificant and positive, as are many of the other HRM interactions reported in Table 3.5. The sole exception is the Employee Participation in Financial Returns × Downsizing interaction (EPFR × DOWNSIZE). Three out of the six interactions are positive in the preferred, right-hand specification with the control variables, and one of these is significant at the 0.10 level. This supports neither the claims-reporting moral hazard nor the risk-bearing moral hazard explanation of why downsizing increases workers' compensation claim durations.

However, our tests for claims-reporting and risk-bearing moral hazard may suffer from specification bias: the functional form may not be the correct one, or there may be omitted factors not included in the claim duration and claim frequency regressions. So in this chapter we present two additional tests for detecting moral hazard effects that have been used previously in the workers' compensation literature. We examine whether claim denial rates are affected by HRM practices, and we examine whether the type of claims (i.e., hard-to-monitor claims versus easy-to-monitor claims) varies with HRM practices.

If, for example, more financial involvement in the firm reduces claim frequency by a reduction in moral hazard, then we would expect that more employee financial involvement would also reduce the number of claim denials on the part of firms. Similarly, since claim-reporting moral hazard is most likely to occur in difficult-to-monitor claims such as back sprains and strains, we would expect that firms offering more financial rewards linked to workers' productivity would experience relatively fewer back strain claims as the moral hazard–induced reporting of those claims falls. We examine the empirical evidence associated with these predictions in the next two sections.

CLAIM DENIAL RESULTS

Our hypothesis is that claims seen by the insurer, or firm, as potential claims-reporting moral hazard behavior are more likely to be denied. We expect that the implementation of safety strategies and worker involvement will ameliorate concerns over claims-reporting moral hazard. Hence, in this section we estimate the following logistic claim denial regression:

$$(4.1) \quad \log\left(\frac{claim\ denial}{1 - claim\ denial}\right) = X\beta + M\delta + H\gamma + D\alpha$$

where *claim denial* is the probability that a given claim will be denied by the firm, M represents Management Safety Culture, H represents other HRM variables, D represents Downsizing, and X represents the remaining control variables in the analysis.

The descriptive statistics for the claim denial analysis are given in Table 4.1 and are the same as those given in Table 3.3 in the last chapter. The logistic regressions in Table 4.2 compute the likelihood that a workers' compensation claim will be denied. Here we will discuss only the results in the right-hand column since this value is the specification that includes the industry, occupational, injury, and year dummy variables. The difference in the specification with and without the control variables is modest: the coefficients are generally of the same sign and magnitude but often lose some significance in the presence of other control variables.

Among the claimant characteristics, only Maleness (MALE) and the Wage Replacement Rate (LnRRATE) have significant impacts: being male increases the likelihood of having one's claim denied by 3.5 percentage points, and increasing the replacement rate by 10 percent increases the likelihood of having one's claim denied by 2.3 percentage points (using the right-hand column estimates). This latter benefits effect is an expected moral hazard result: the higher the replacement rate, the greater the incentive to report a claim and the more likely that it will be denied. Downsizing in the firm also has the expected positive coefficient but is statistically insignificant in either specification in Table 4.2. The insignificant coefficient for Downsizing in the claims denial regression suggests that it may not be a good proxy for claims-reporting (or risk-bearing) moral hazard in this sample. This insignificance also suggests that our interpretation of the interaction effects in the last section as indicators of moral hazard response is probably not a very discerning test.

Nonetheless, if firms perceive that there is a significant amount of claims-reporting moral hazard taking place, then the existence of HRM practices that increase job satisfaction or a worker's financial commitment to the firm ought to be associated with lower claim denial rates (Card and McCall 1996). That is, the HRM practice variables ought to have negative coefficients. However, the HRM coefficients offer mixed results. Only three of the five coefficients are negative. Moreover, in the preferred specification given in the right-hand column of Table 4.2 only the financial returns variable (EPFR) is statistically significant at the 0.10 (but not the 0.05) level. This suggests that the reductions in claim severity due to HRM practices are not exclusively the result of reduced moral hazard. If they were, then more of these HRM practices would be associated with fewer claim denials.[3]

Table 4.1 Variable Definitions and Descriptive Statistics

Variable	Definition	Mean	Std. dev.
Dependent variables			
NWSPELL	Duration of nonwork spell: days of temporary total disability benefits paid in the workers' compensation system[a]	55.78	140.91
DENIAL	A dummy variable coded 1 if the claim was denied for the liability by the insurer, 0 otherwise	0.20	0.40
Downsizing variable			
DOWNSIZE	A dummy variable coded 1 if the claimant's company experienced layoffs or cutbacks in the last 12 months, 0 otherwise		
HRM variables			
EPDM	Number of employee participation programs in decision-making process	3.12	1.36
EPFR	Number of employee participation programs in the firm's financial returns	3.08	2.16
MGTCULT	Score of management's commitment to workplace safety	22.62	2.29
EPDEG	Score of the degree of employee participation in the company's decision-making process	8.12	1.00
INFOSHR	Score of the degree to which management shares information with employees on the various issues in the production process	20.26	5.24
Claimant characteristics			

Variable	Description		
MALE	A dummy variable coded 1 if claimant is male, 0 if female	0.53	0.50
LnAGE	Log of age of claimant at time of injury	3.63 (39.47)[b]	0.31 (11.54)[b]
LnRRATE	Log of wage replacement rate in Minnesota workers' compensation system[c]	4.28 (73.40)[b]	0.19 (14.41)[b]
SELF	A dummy variable coded 1 if claimant's company is self-insured for its workers' compensation coverage	0.29	0.45

[a] Descriptive statistics of variables without taking logs.

[b] The dependent variable NWSPELL is transformed into logarithmic form in the LIFEREG function of SAS.

[c] Real wage replacement rate was used to capture both wage and expected workers' compensation benefit effects on the dependent variable. In accordance with the Minnesota workers' compensation law, RATE was calculated by the following formula (Minnesota WC income benefit schedule used; 1992 analysis of workers' compensation laws, U.S. Chamber of Commerce):

$$\text{RATE} = \text{MAX}_t / \text{Wage} \quad \text{if } (\text{Wage} \times 0.66) \geq \text{MAX}$$
$$0.66 \quad \text{if } [\text{MIN}_t \leq (\text{Wage} \times 0.66) < \text{MAX}]$$
$$\text{MIN}_t / \text{Wage} \quad \text{if } [(\text{MIN}_t \times 0.66) \leq (\text{Wage} \times 0.66) < \text{MIN}_t]$$
$$1 \text{ otherwise,}$$

where Wage is average production employee's gross weekly wage, MAX, is maximum amount of wage replacement through Minnesota workers' compensation system, and MIN, is minimum amount of wage replacement through Minnesota workers' compensation system. (In the analysis of this study, log of RATE × 100 was included in the models as RRATE).

SOURCE: Authors' calculations.

Table 4.2   Logit Estimates of Claim Denials in the Workers' Compensation System (standard error in parentheses)

Variable	Dependent variable: incidence of claim denial in the Minnesota workers' compensation system	
	Model (1)	Model (2)
Constant	−8.87***	−8.39***
	(1.49)	(1.78)
Downsizing variable		
DOWNSIZE	0.17	0.09
	(0.11)	(0.13)
HRM variables		
EPDM	0.15***	0.09*
	(0.04)	(0.05)
EPFR	−0.15***	−0.10*
	(0.04)	(0.05)
MGTCULT	−0.03	0.05
	(0.04)	(0.05)
EPDEG	0.06	−0.08
	(0.07)	(0.08)
INFOSHR	−0.01	−0.004
	(0.01)	(0.01)
Claimant characteristics		
MALE	0.14	0.22*
	(0.10)	(0.12)
LnAGE	0.38***	0.22
	(0.15)	(0.16)
LnRRATE	1.45***	1.55***
	(0.26)	(0.29)
SELF	−0.49***	−0.26
	(0.11)	(0.16)
Control variables		
Industry dummies	no	yes
Injury type dummies	no	yes
Occupation dummies	no	yes
Year dummies	no	yes
−2 log likelihood	3226.74	2858.64
n	3356	3104

NOTE: * significant at the 0.10 level (two-tailed test); ** significant at the 0.05 level (two-tailed test); *** significant at the 0.01 level (two-tailed test).
SOURCE: Authors' calculations.

CLAIM TYPES: MULTINOMIAL LOGIT ANALYSIS

If HRM practices lower moral hazard behavior, HRM practices ought to have an impact on the type of claims that are filed. If the potential for moral hazard is greatest where the costs of monitoring are largest, then relatively more evidence of moral hazard should be seen with respect to those types of claims that are most difficult to monitor. Again, this is especially true for sprains, strains, and cumulative trauma conditions.

Detecting moral hazard incentives by comparing difficult-to-monitor injuries with easy-to-monitor injuries was previously employed by Smith (1990) and by Dionne and St-Michel (1991). In analyzing the times when claims have been filed, Smith raises the possibility that workers' compensation may be paying for some off-the-job injuries. Smith argues that off-the-job injuries reported as work-related would probably be difficult to diagnose, relatively easy to conceal, and would be reported early in the shift, especially on Mondays (the so-called Monday morning syndrome). He finds that of the three largest categories of claims, sprains and strains are reported earlier in the day. Moreover, the propensity to report sprains and strains earlier in the day is significantly increased on Mondays and Tuesdays following a three-day weekend. Smith estimates that 4 percent of sprains and strains are misrepresented as having occurred on the job. Card and McCall (1996) don't dispute Smith's findings in their analysis of Minnesota data, but they discount the moral hazard interpretation of the findings since employers did not increase claim denials for those claims filed on Monday, as would be expected if the additional filings were related to claims-reporting moral hazard.

Dionne and St-Michel (1991) examine moral hazard by looking at the variation in days on workers' compensation for those with difficult-to-diagnose conditions compared to those with less-difficult-to-diagnose conditions. They partition injuries into two categories based on injury severity (minor injuries with fewer lost days or major injuries with greater lost days) and whether the condition is easy or more difficult to diagnose. Like Smith (1990), they reason that moral hazard response will be greatest for the difficult-to-diagnose injuries: lower back pain (minor injury) and spinal disorder (major injury). They find that as Quebec's coinsurance rates decreased in 1979, days workers spent off work on a difficult-to-diagnose claim rose significantly more than did days from claims falling into the easy-to-diagnose category. They also found that, once the interaction with diagnostic difficulty was controlled for, the 1979 shift in coinsurance rates had no independent effect on days on a claim. That is, most of the impact the declining coinsurance rates had on days on a claim came through an increase in days consumed by those with difficult-to-diagnose injuries.

Butler, Durbin, and Helvacian (1996) use this distinction between difficult-to-monitor and easy-to-monitor injuries to explore whether soft-tissue injury claims correlate with level of benefits and spread of HMOs. They find in their 10-year, 15-state sample of workers' compensation claims that the proportion of claims attributable to soft-tissue injuries rose from 44.7 percent of all claims in 1980 to 50.6 percent in 1989. Concurrently, the share of costs attributable to soft-tissue injuries rose from 41 percent to 48.8 percent. The share of costs for injuries that crush or fracture a bone—easy-to-monitor claims—is the only category that declined between 1980 and 1989. Using a multinomial logit model, the authors determine that most of the increase in soft-tissue injury is attributable to the expansion of HMOs. Specifically, they ascribe the rise in such injuries to moral hazard response by HMO providers, who increase their revenue by classifying as work-related injuries as many health conditions as possible.[4]

We build on the work done by Butler, Durbin, and Helvacian (1996) by considering whether HRM practices affect the distribution of injuries. Given prior evidence on soft-tissue sprain and strain, we would expect to see additional HRM practices associated with fewer sprains and strains (particularly back sprains and strains), and with relatively more fractures and lacerations, *if* HRM practices are reducing workplace injuries through a claims-reporting response.

The effect of Management Safety Culture on risk-bearing moral hazard is not as clear as the effect of other HRM practices on claims-reporting moral hazard. An increase in risk-bearing moral hazard may increase any of the claim categories, so the change in injury types that is attributable to a reduction in risk-bearing moral hazard is ambiguous. Hence, the Management Safety Culture variable, which of all the HRM variables most directly impacts the degree of risk-bearing moral hazard, will have an ambiguous sign across the four injury groups.

On the other hand, if HRM practices work to reduce claims-reporting moral hazard, we would expect to find them reducing the proportion of back sprains and strains and increasing the relative number of fractures, contusions, and cuts.

To place the ideas discussed above into a statistical framework, we assume the typical worker experiences one of five states. The worker may not have any type of health impairment whatsoever, or the worker may experience some sort of injury that places him into one of four injury categories: 1) fractures, contusions, and cuts; 2) back sprains and strains; 3) nonback sprains and strains; and 4) all others.

Conditional on gender, age, the insurance benefit replacement rate, and whether the firm is self-insured, a probability distribution describes the likelihood of being in each of these five states. Because of moral hazard response, the marginal worker will migrate from one state to another when there is an incentive to do so. In particular, as financial rewards increase through profit sharing and other forms of financial returns, or job satisfaction increases through greater involvement with firm decision making, then the utility-maximizing worker is less likely to migrate to the difficult-to-diagnose (and easy-to-feign) category of back sprains and strains. The HRM effects on back sprains and strains will be negative, and the HRM effects on fractures will be positive (or at least, more positive than they are for lower back sprains and strains).

Since data are available only on workers who report claims, the noninjured state is omitted. The stochastic specification employed below implies that the parameter estimates will be unchanged by such an omission; the odds ratio implied by a multinomial logit model maintains the independence of irrelevant alternatives, thus the parameter estimates will be consistent.[5] The categorical dependent variable identifies one of the four groups of injuries above. Although the parameters for other sprains and strains (nonback) are necessarily normalized, the implied impact of the HRM variables on nonback sprains and strains is given in Table 4.4.

The multinomial logit model used in this analysis assumes that the injured worker's perceived wellness, or utility, is given by

$$(4.2) \quad U_{ij} = U(W_{ij}, S_{ij}) = X_i \beta_j + \varepsilon_{ij},$$

where W_{ij} represents financial incentives of the worker and S_{ij} represents nonwage dimensions of work (i.e., the worker's involvement with firm decision making) of worker i in claimant status j. The vector X_i includes factors that determine financial and nonfinancial aspects of the job and is assumed to be constant across claimant states. The vector of coefficients, β_j, and the random error term, ε_{ij}, vary by claimant status. Note that this is a more general model than is typically employed in the literature, where only utility in the injured versus noninjured states is considered.

Consider the marginal worker who chooses to file a back sprain or strain claim. The worker follows this pattern because he derives more utility from it than from any of the other alternatives: presumably the other claim types do not yield as much utility given the firm's financial and nonfinancial dimensions, and the utility he gets while on a sprain or strain claim (which we will denote as state "1") is higher than it would be if he worked (state "W," with the other injury states labeled 2, 3, and 4). Thus, the probability of observing a worker on a sprain or strain claim is

(4.3) $P_{i1} = Pr\left[(U_{i1} > U_{iw}) \cup (U_{i1} > U_{i2}) \cup (U_{i1} > U_{i3}) \cup (U_{i1} > U_{i4})\right]$

$= Pr\left[(X_i\beta_1 - X_i\beta_w > \varepsilon_{iw} - \varepsilon_{i1})\right.$

$\cup (X_i\beta_1 - X_i\beta_2 > \varepsilon_{i2} - \varepsilon_{i1})$

$\cup (X_i\beta_1 - X_i\beta_3 > \varepsilon_{i3} - \varepsilon_{i1})$

$\left. \cup (X_i\beta_1 - X_i\beta_4 > \varepsilon_{i4} - \varepsilon_{i1})\right].$

Similar expressions hold for the other three claimant states.

Assuming that the underlying distributions of ε_{ij} are type I extreme-value-distributed, the probability of observing the ith worker in claimant status j is

(4.4) $P_{ij} = \dfrac{\exp(X_i\beta_j)}{1 + \sum_{k=1}^{3} \exp(X_i\beta_k)}$

(Maddala 1983).

We define a set of four dummy variables such that

$Y_{ij} = 1$ if the ith worker files claim type j ;
$Y_{ij} = 0$ otherwise.

Then, the log-likelihood function for the model is

$$(4.5) \quad \log \ell = \sum_{i=1}^{n} \sum_{j=1}^{4} Y_{ij} \log P_{ij} .$$

Maximizing this equation yields estimates of the parameters β_j for the three claim types relative to the omitted claim category, "All other claims."

Note that the coefficients of the multinomial logit function do not represent the marginal effects of the independent variables on claim choice. However, the coefficients may be converted to measures of marginal effects and, for ease of comparison among the variables, expressed as semi-elasticities using well-established formulas:

$$(4.6a) \quad \frac{\partial P_j / P_j}{\partial P_j} = \beta_{mj} - \sum_{k=1}^{3} \beta_{mk} P_k \quad \text{for } j = 1,2,3 ;$$

$$(4.6b) \quad \frac{\partial P_j / P_j}{\partial X_j} = - \sum_{k=1}^{3} \beta_{mk} P_k \quad \text{for } j = 4 .$$

In Equations (4.6a) and (4.6b), k and j are indices across the four claim types, while m is an index across the explanatory variables in X_i.[6] The control variables in this analysis are gender, age, the worker replacement rate, and whether the worker's firm self-insures.

Table 4.3 presents the sample means for the 3,104 claimants employed by firms in our sample data: 11 percent are fractures, contusion, or cuts; 21 percent are back sprains or strains; 18 percent are other sprains or strains; and 49 percent are all other injury types. The means for the independent variables are about the same as those in Table 3.3 of the last chapter. This claimant sample has slightly more males than females, an average age of 40, relatively low reported average weekly wages, and a relatively high replacement rate. Across the injury types, the claimants show little variation in HRM practices among their firms except for Employee Participation in Financial Returns (EPFR) and Information Sharing (INFOSHR), where the standard deviation is signifi-

Table 4.3 Means of Independent Variables for Full Sample and by Injury Type (standard deviation in parentheses)

Variable	Total sample (3,104)	Injury types			
		Fracture, contusion, and cuts (345)	Back sprain and strain (667)	Other sprain and strain (557)	All other types of injuries (1,535)
Downsizing variable					
DOWNSIZE	0.424 (0.494)	0.383 (0.487)	0.351 (0.478)	0.429 (0.465)	0.463 (0.499)
HRM variables					
EPDM	3.350 (1.376)	3.116 (1.350)	3.349 (1.460)	3.232 (1.308)	3.447 (1.360)
EPFR	3.241 (2.291)	3.107 (2.142)	2.751 (2.004)	3.303 (2.358)	3.461 (2.381)
MGTCULT	22.877 (1.706)	22.986 (1.615)	23.099 (1.780)	22.774 (1.760)	22.793 (1.664)
EPDEG	8.232 (1.005)	8.157 (1.005)	8.084 (1.089)	8.201 (1.087)	8.325 (0.924)
INFOSHR	21.423 (4.838)	21.128 (4.991)	20.640 (4.716)	21.557 (4.755)	21.782 (4.847)

Control variables

MALE	0.550 (0.497)	0.664 (0.473)	0.504 (0.500)	0.530 (0.500)	0.551 (0.498)
EEAGE	3.634 (0.304)	3.642 (0.324)	3.567 (0.305)	3.598 (0.300)	3.633 (0.297)
RRATE	4.273 (0.178)	4.271 (0.195)	4.278 (0.175)	4.264 (0.167)	4.275 (0.181)
SELF	0.241 (0.428)	0.223 (0.417)	0.375 (0.484)	0.237 (0.426)	0.189 (0.392)

NOTE: See Table 3.3 for data sources and calculations.

[a] Based on parameter estimates reported in Table 4.5.

[b] The standard errors given for the elasticities are asymptotic approximations based on the estimated standard errors in Table 4.5. Therefore, the standard errors for the elasticities are a much less reliable indicator of the statistical significance of a particular variable than are the chi-square tests given in Table 4.5.

SOURCE: Authors' calculations.

cantly larger than it is for the other HRM practices. For both of these HRM practices, claimants with back sprains work in firms with fewer financial returns programs (2.75 for back sprains versus 3.24 across all injury types) and with less information sharing (20.64 for back sprains versus 21.42 across all injury types). This is consistent with the claims-reporting moral hazard information of claims filing: in firms with fewer HRM practices, more hard-to-monitor (lower back sprain) claims are filed. However, these raw descriptive differences do not necessarily measure the influence of HRM practices: other confounding variables may be simultaneously influencing injury types.

To control for these other potentially confounding variables, we use a multinomial logit model in which the type of claim filed is made a function of the same variables used in the analysis of claim duration in the last chapter. The multinomial logit estimates of the first three claim types in Table 4.3, relative to the "other sprains and strains" category, are reported in Table 4.5 for 3,104 claims for which we have complete data on the HRM practice, demographic, and occupational and industry control variables. The chi-square statistics in Table 4.5 indicate that the Downsizing variable and two of the HRM practice variables, Employee Participation in Financial Returns (EPFR) and Information Sharing (INFOSHR), are statistically insignificant, while the degree and intensity of Employee Participation in Decision Making (EPDEG and EPDM) and Management Safety Culture (MGTCULT) variables are significant.

Since the multinomial logit coefficients do not represent the marginal effects of different characteristics on the probability of making a particular type of claim, we transform the coefficients to partial derivatives and, for ease of comparison, express the coefficients as semi-elasticities. These semi-elasticities, with their approximate standard errors in parentheses, are reported in Table 4.4. The semi-elasticities provide only weak evidence of claims-reporting moral hazard. Consistent with the explanation that HRM practices lower workers' compensation costs by decreasing claims-reporting hazard, an increase in any of the four HRM practices (except Management Safety Culture) lowers the likelihood of a lower back sprain claim being filed. For example, adding another program that involves the worker in firm decision making (EPDM increases by 1) will lower the likelihood of a lower back sprain claim being filed by 1.5 percent. Adding another program that increases the worker's involvement in the financial returns of the company lowers the

Table 4.4 Semi-Elasticity Estimates of the Effects of Human Resource Policies on the Type of Injury[a]: Multinomial Logit Regression[b] (standard error in parentheses)

| | Injury types | | | |
Variable	Fracture, contusion, and cuts	Back sprain and strain	Other sprain and strain	All other types of injuries
Intercept	−4.877	5.301	2.573	−2.139
	(1.36)	(1.11)	(.14)	(1.00)
Downsizing variable				
DOWNSIZE	−.015	−.0147	−.017	.106
	(.28)	(.28)	(.30)	(.15)
HRM variables				
EPDM	−.150	−.015	−.134	.089
	(.09)	(.10)	(.13)	(.10)
EPFR	.051	−.011	.074	−.034
	(.10)	(.11)	(.12)	(.14)
MGTCULT	.141	.015	.025	−.047
	(1.34)	(1.11)	(.32)	(.96)
EPDEG	−.225	−.068	−.123	.125
	(.42)	(.39)	(.37)	(.29)
INFOSHR	.020	−.007	.009	−.005
	(.30)	(.26)	(.31)	(.32)
Control variables				
MALE	.274	.107	−.117	−.066
	(.14)	(.16)	(.12)	(.12)
EEAGE	.503	−.592	−.219	.223
	(.23)	(.20)	(.12)	(.18)
RRATE	.170	−.801	−.396	.453
	(.12)	(.09)	(.08)	(.14)
SELF	.071	.418	−.020	−.190
	(.10)	(.09)	(.09)	(.08)

NOTE: The specification also includes dummy variables for years, industry, and workers' occupation, but the corresponding coefficients are not reported here. While none of these control variables were statistically significant individually, log-likelihood ratio tests indicate that they were jointly significant at greater than the 0.01 level.
[a] Based on parameter estimates reported in Table 4.5.
[b] The standard errors given for the elasticities are asymptotic approximations based on the estimated standard errors in Table 4.5. Therefore, the standard errors for the elasticities are a much less reliable indicator of the statistical significance of a particular variable than are the chi-square tests given in Table 4.5.
SOURCE: Authors' calculations.

Table 4.5 The Effects of Human Resource Policies on the Type of Injury: Multinomial Logit Regression (standard error in parentheses)

	Injury types[a]			
Variable	Fracture, contusion, and cuts	Back sprain and strain	All other types of injuries	$x^{2\,b}$
Intercept	−7.450*** (2.62)	2.730 (2.16)	−4.712** (1.92)	24.90***
Downsizing variable				
DOWNSIZE	.002 (.19)	.002 (.16)	.032 (.14)	.09
HRM variables				
EPDM	−.015 (.07)	.119** (.06)	.223*** (.05)	24.61***
EPFR	−.023 (.08)	−.085 (.07)	−.108* (.06)	4.10
MGTCULT	.116* (.07)	−.010 (.06)	−.072 (.05)	10.33**
EPDEG	−.102 (.11)	.055 (.09)	.248*** (.08)	17.78***
INFOSHR	.011 (.02)	−.015 (.02)	−.014 (.01)	2.85
Control variables				
MALE	.391** (.18)	.224 (.15)	.051 (.13)	6.49*
EEAGE	.721*** (.24)	−.373* (.20)	.442** (.17)	34.14***
RRATE	.566 (.45)	−.406 (.37)	.849*** (.32)	18.99***
SELF	.090 (.24)	.437** (.19)	−.171 (.17)	15.70***

NOTE: Total number of injuries was 3,104; −2 log likelihood = 7303.86. * significant at the 0.10 level (two-tailed test); ** significant at the 0.05 level (two-tailed test); *** significant at the 0.01 level (two-tailed test).

The specification also includes dummy variables for years, industry, and workers' occupation, but the corresponding coefficients are not reported here. While none of these control variables were statistically significant individually, log-likelihood ratio tests indicate that they were jointly significant at better than the 0.01 level.

[a] The omitted injury type is "Other sprain and strain."

[b] Chi-square statistics for each variable.

SOURCE: Authors' calculations.

likelihood of a lower back sprain claim being filed by 1.1 percent. The moral hazard explanation implies an ambiguous impact for the Management Safety Culture coefficient, and it is the only HRM practice that is not negative.

However, the evidence that HRM practices reduce claims-reporting moral hazard is ambiguous for two reasons. First, HRM effects on lower back sprains are empirically small (i.e., the semi-elasticities are small in magnitude). While the EPDM and EPDEG are statistically significant (see Table 4.5 in the right-hand column for the test of joint significance), the switch away from the potentially moral-hazard-laden lower back sprain category as these programs increase is too small to account for the magnitude of effect reflected in the HRM reduction of workers' compensation costs in Tables 3.2, 3.4, and 3.6 of Chapter 3.

Second, the multinomial estimates provide no corroborative evidence of moral hazard behavior with regard to other (non-HR) results: Downsizing has no impact on claim types, increases in the replacement rate do not increase the proportion of lower back sprains, and self-insurance does not lower the proportion of lower back sprains. As the replacement rate increases, the opportunity cost of being out of work on a workers' compensation claim falls. Claims-reporting moral hazard will likely increase, especially for injuries (such as lower back sprains) whose work origin is difficult to monitor or detect. Hence, an increase in claims-reporting moral hazard ought to increase the proportion of low back sprains.

Similarly, firms that self-insure their workers' compensation claims have the greatest incentive to monitor claims for moral hazard behavior since they bear the full cost of such behavior, whereas firms that are not fully experience-rated do not. For this reason, if claims-reporting moral hazard were significant, we would expect those firms that self-insure to have relatively fewer lower back claims. However, proportionately fewer lower back claims are filed for those firms that have higher replacement rates or that do not self-insure. Hence, the Downsizing (DOWNSIZE), Benefit Replacement Rate (RRATE), and Firm's Self-Insured Status (SELF) coefficients have the opposite impact on the filing of lower back sprains if the filing of lower back sprains mostly reflects moral hazard behavior.

CONCLUSIONS

In Chapter 1, we noted that HRM practices can affect any one of the three dimensions of reported injury loss. First, they can affect the propensity to file a claim, even if the intrinsic workplace risk remains unchanged. Workers' reporting propensities are affected by the degree of insurance coverage, so there is a claims-reporting moral hazard. Intrinsic risk is the level of physical danger of accidental injury or occupational disease that is the result of workers' producing output, and injury costs may be reduced by modifying either or both of the following: by changing workers' incentives to take care (through changes in risk-bearing moral hazard), or by modifying the ergonomic aspects of the physical workplace, including employing processes, procedures, or equipment that reduce on-the-job injuries.

We have given three explanations for our findings of the last two chapters of why HRM practices reduce workers' compensation costs:

1) Claims-reporting moral hazard changes with HRM practices,

2) Risk-bearing moral hazard changes with HRM practices, and

3) Intrinsic risk improves (independent of employees' change in safety behavior) with HRM practices.

Our evidence is more supportive of the latter two explanations than it is of the first.

Claims-reporting moral hazard. Our findings do not consistently support a claims-reporting moral hazard explanation of why HRM practices reduce workers' compensation costs. We find no strong evidence that the Downsizing effect represents a moral hazard effect, either of the risk-bearing or claims-reporting type. Thus it is tenuous to view Downsizing or HRM practice interactions as providing tests for moral hazard behavior. This is the reason that we also analyzed the impact of HRM practices on claim denials and claim types in this chapter.

The analysis of this chapter provides little evidence that HRM practices operate solely, or mostly, through a reduction in claims-reporting moral hazard. The only HRM practice that appears to significantly reduce the likelihood of a claim denial is Employee Participation in Financial Returns, which, although consistently exhibiting the expected sign in all of the analyses in this book, remains a relatively small effect. Employee Participation in Financial Returns is marginally significant in the claims denial analysis; Employee Participation in Financial Returns is marginally insignificant in the multinomial logit analysis.

All HRM practices do have the expected impact on the filing of hard-to-monitor claims (such as lower back sprains and strains) in the multinomial logit regression. This is evidence of claims-reporting moral hazard. But the estimated effect is small, too small to explain why these programs reduce workers' compensation costs as much as they appear to do. Moreover, the coefficients of other ancillary variables that would also indicate the presence of claims-reporting moral hazard—DOWNSIZE, RRATE, and SELF—do not have the expected sign.

This doesn't mean that moral hazard isn't important, only that the claims denial and multinomial logit results indicate that HRM practices do not reduce workers' compensation solely, or even mainly, through reductions in claims-reporting moral hazard response.

Risk-bearing moral hazard. The results are also consistent with changes in risk-bearing moral hazard. But a risk-bearing moral hazard explanation of the HRM impact accommodates the claim denial results more readily than does a claims-reporting explanation. With a change in the number of real injuries (as might come through a reduction in risk-bearing moral hazard), there would be no change in claim denials on the basis of employee claims-reporting moral hazard since there would be no grounds on which to contest the validity of these claims. Moreover, changes in workers' risk-bearing activities would have an ambiguous impact on the types of claims filed, hence HRM practices would have an indeterminate effect on the distribution of claims even if the HRM practices served to change the degree of risk-bearing behavior on the part of workers.

Change in real safety, independent of any change in workers' safety behavior. Besides a change in risk as workers change their risk-taking behaviors, there can be changes in risk brought about by the introduction of better safety information, better safety practices, better safety equipment, and better allocation of labor to heterogeneous tasks even if workers don't change their behavior because of insurance coverage. Involvement of workers in strategic safety planning or in the financial returns of the firm leads to more efficient safety outcomes if workers are the least-cost provider of safety information and safety processes. This explanation would also be consistent with the results reported in this book. It seems to us to be a plausible research lead to pursue.

Notes

1. Benefits replace only two-thirds of lost wages, subject to a maximum benefit (which restricts most potential recipients' income to less than two-thirds) and a waiting period (three days in Minnesota) during which no benefits are received. Moreover, there are no benefits paid for pain and suffering.

2. Claim duration may rise or fall, depending on whether the change in claim frequency substantially shifts the relative number of short- and long-duration claims. If this composition effect is small, we would expect the average claim duration to increase with downsizing as injury severity increases with injury risk.

3. Logistic regressions for claim denials with Downsizing interaction variables were also estimated, analogously to the duration estimates in Table 3.3 of the last chapter. However, none of the interactions were statistically significant at the 0.05 level (and none at all in the full model with the control variables). These results are not reported here.

4. Butler, Hartwig, and Gardner (1997) find, in a panel data set of selected states followed during the 1980s, that HMOs indeed have a greater tendency to classify claims as compensable under workers' compensation than do independent physicians. Their findings suggest that real workers' compensation costs might have declined during the period except for the rapid expansion of HMOs and the perverse incentives generated by the potential dual coverage (i.e., workers' compensation or the per-capitated plan) of various health conditions.

5. See the discussion in Maddala (1983), especially p. 77.

6. Standard errors for the semi-elasticity estimates are computed by taking square roots of the diagonal elements of the matrix ZSZ', where $S = \text{Cov}(\beta_{ij})$ and

$$
Z = \begin{bmatrix} 1-P_1 & -P_2 & -P_3 \\ -P_1 & 1-P_2 & -P_3 \\ -P_1 & -P_2 & 1-P_3 \\ -P_1 & -P_2 & -P_3 \end{bmatrix} \otimes I_k.
$$

5
How Much Safety Is Desirable?

CAUTIONS AND OUR EMPIRICAL FINDINGS

Our results are conditioned on the sample we employed: medium-sized and small firms. Most of the firms were concerned with safety. In this regard, we believe that our sample is representative of small-to-medium-sized firms. Our survey response rate was relatively high. Even though our survey instrument measured more dimensions of safety management than any previous study, there still may be omitted factors—factors for which we do not have any controls. This is a possible shortcoming in every study that does not employ completely random assignment, and one we hope we have minimized here by employing a rather extensive list of firm management and safety factors. And while our sample consists of those firms that applied to the Minnesota Safety Grant Program, we have no reason to believe that these firms are substantially different from firms in general. Nevertheless, our results are strictly valid for only those firms included in our sample.

Job requirements, both for the firms in our sample and for the United States, have become increasingly specialized. This specialization has taken place not only in the levels of technical know-how required but in the organizational tools specific to each firm. Loss of skilled labor in the competitive marketplace can be especially costly, as it often takes two to six months to train new workers. Firms that can retain their skilled workers in productive employment have a competitive advantage over those that cannot. While there has been a lot written on the importance of maintaining skilled employees by reducing job leaving, it is rarely noted that firm-specific human capital is also lost when employees temporarily leave because of workers' compensation claims.

In this study we analyze the impact of various human resource management practices on a firm's workers' compensation costs. We partitioned HRM practices into two groups: 1) practices that the firm can unilaterally adopt that do not necessarily involve the workers in either the financial or the strategic management of the firm, and 2) practices that increase workers' involvement with the firm in one or both of these dimensions. Among the former, known as management environmental factors, we include Management Safety Culture (MGTCULT, the degree of management involvement with the safety efforts of the firm) and Information Sharing (INFOSHR) as variables. Among the latter group of HRM practices, those with explicit worker involvement, we have three variables to approximate different dimensions of employee participation: 1) the number of programs that allow the employee to participate in the financial returns of the company (EPFR), 2) the number of programs that allow the employee to participate in the strategic planning of the company (EPDM), and 3) the intensity of the employee's involvement in the strategic planning of the company (EPDEG).

Of these five dimensions, we find that greater information sharing by management and greater intensity in the firm's strategic planning by employees raises rather than lowers workers' compensation costs. Increased information sharing by management with employees increases claim frequency by as much as it decreases claim duration, so the net effect is a small and insignificantly positive increase in costs. The insignificant net impact on costs is consistent with the results in Chapter 4: greater information sharing doesn't increase the likelihood of claim denials, as would be expected if there were substantial claims-reporting moral hazard, nor is it associated with a shift in the distribution of injury types, as would be expected if there were risk-bearing moral hazard present.

On the other hand, the intensity of employee involvement has a positive net impact on workers' compensation costs as a result of a 19 percent increase in the duration of a claim for each unit increase in the intensity index. Since this index measures intensity of worker involvement on a scale of 2 to 10, with higher scores indicating that workers feel they "always" have control over their job tasks and "always" participate in employee involvement programs, we had hypothesized that more intense involvement would lead to lower costs. The opposite appears to be true. In the samples, the mean intensity scores were around 8, indicating that employees participate rather intensely in the activities at their firm. There are at least two explanations for this unexpected result of higher costs: 1) a greater sense of privilege when hurt, and 2) more risk taking. The former explanation hinges on a sense of ownership—meaning a greater sense of entitlement—when an employee gets injured, so that he feels justified in taking more time to recover from his work injury. The latter explanation has to do with the employee feeling that he has more control over his job than he had in the past. Because of this feeling of control, he is willing to take more job risks that result in more serious injuries. The latter explanation is consistent with the results reported in Chapter 4: larger values of the Intensity of Employee Participation (EPDEG) don't increase the likelihood of claim denials, but they do significantly shift the distribution of injury claims toward the "all other" category, which tends to have longer claim duration.

Increases in any of our three main HRM practices—EPDM, Employee Participation in Decision Making; EPFR, Employee Participation in Financial Returns; and MGTCULT, the level of management involvement in the safety processes of the firm—all lead to substantial reductions in workers' compensation costs per employee. In Table 3.6, we measured per-employee safety gains both as a unit change in each of these indices and as a change to the "best practice" levels.

As an example of a unit change interpretation of our results, we find that if a firm were to go from using two to using three financial return programs (EPFR went from 2 to 3), then claim frequency would fall by 23 percent and claim duration would fall by 13 percent. Lost work-time benefit costs would fall by 36 percent (the combined impact of frequency and duration outcomes), which when multiplied by the average indemnity cost per employee of $175.50 yields a cost reduction in lost-time expenses of $63.18 per employee per year. This is the approximate reduction in per-employee costs of adding another plan in which the employee shares in the financial returns of the company—that is, of making a unit change in the HRM practices.

There is another interpretation of the results: suppose that the firm were to go from an average, 50th percentile firm to an 84th percentile firm, i.e., to one of the top 16 percent of firms. If "best practice" firms were those with the highest 16 percent of EPFR values, what would be the change in costs of moving from the 50th percentile to the 84th percentile? This change is a one-standard-deviation change in the value of the EPFR; in this case, the standard deviation would be a change of approximately 1.211 more programs. (In such a case, a standard deviation change is small because there is little variation across our sample firms in the number of financial return programs they offer to employees.)

When we add one standard deviation more of EPFR programs to their current offerings, then costs fall by 1.211 × $63.18, or by $76.51.

The best-practice interpretation of our findings is probably the most relevant, given the sample variation in these alternative HRM practices across firms. As indicated by the far right-hand column in Table 3.6, an average firm that becomes a best-practice firm in EPDM saves about $36 per employee per year, an important level of savings but only half of the best-practice savings available from EPFR.

The HRM practice that represents the largest cost savings per employee is Management Safety Culture. A firm that increases its level of involvement from the average to the best-practice level of MGTCULT, a variable increase of about 3, saves about $180 per year per employee. In a firm with 100 employees, this is an annual savings of $18,000 in lost-time pay alone. This is likely to be a lower bound on the benefits derived from improving management safety culture. In addition to the lost-time pay costs, medical costs will be saved, and employees' level of job satisfaction may well increase as well. Uncompensated wage loss and pain and suffering associated with injuries will fall, reducing the compensating wages paid to workers. Perhaps equally important in our highly skilled labor market, more firm-specific human capital can be retained as the number of days out due to injuries falls.

Since these are potentially significant savings, it is useful to understand which one of three channels they are achieved through: 1) claim reductions due to less claims-reporting moral hazard, 2) claim reductions due to less risk-bearing moral hazard, or 3) claim reductions due to more efficient use of safety resources. The empirical models in Chapter 4 were an attempt to sort out these alternative explanations. We find, as is consistent with prior research, evidence of claims-reporting moral hazard with respect to increases in the replacement rate: claim duration and claim denials increase with higher benefit replacement ratios. However, we find no strong evidence that a higher level of any of the HRM practice variables changes employee behavior, whether in regard to employees' willingness to report claims or employees' willingness to bear on-the-job risk. There is some evidence that EPFR is consistent with a reduction in risk-bearing moral hazard, but the magnitude of the effect is too small to account for the relatively large EPFR effect on costs. Others should pursue these questions with larger samples from other states, but in the meantime, we are left to conclude that the HRM practice effects operate mainly through the third channel: the more management and worker involvement there is with the safety processes of the firm, the safer the workplace becomes. Safety outcomes improve as safety resources are used more efficiently.[1]

IMPLICATIONS FOR FIRMS

The most important HRM best practice was found to be increasing management involvement with the safety programs of the firm. The measured safety gains associated with the MGTCULT variable are substantial, enough so that they are likely to be cost-effective when the costs associated with these activities are taken into account. Recall that MGTCULT is a Likert scale index of responses to the following issues:

1) management's support for clear goals and objectives on safety and health policy,

2) management's leadership in setting goals on safety and health,

3) management's interest in safety and health issues as a part of the firm's strategic level of decision making,

4) management's willingness to share safety-related information with employees, and

5) management's commitment to reemployment of disabled workers and to having a return-to-work program for injured employees.

None of the first four characteristics of good management safety culture listed involve any significant cost to implement. Only the fifth, a commitment to a proactive return-to-work program, may be costly. While evidence on the costs of an effective return-to-work program is lacking, there is some evidence on its benefits. Hunt et al. (1993), in their sample of larger Michigan firms, report that the presence of a proactive return-to-work program significantly lowers workers' compensation claim rates. Butler, Johnson, and Baldwin (1995) find that all three types of job accommodations they measured increase the likelihood of a successful return to work for their sample of severely injured workers: reduced-hour accommodations had the smallest impact, followed by light-work accommodations, and offering modified equipment had the greatest impact on a successful return to work. However, their data did not contain any information on the costs of providing such accommodations. But while the cost-effectiveness of accommodations is not well understood, the evidence certainly suggests that there are some benefits from having a proactive return-to-work policy in place.

Though employee participation programs also lower workers' compensation costs, it is more difficult to interpret how cost-effective they might be, because we have no measure of the firm's cost of implementing these programs. At the very least, involvement of workers with the decision making of the firm involves time away from the production line. This type of forgone output cost may be small, but the firm would have to weigh the implicit loss of output against the benefits we estimate here. We think that calculation will be favorable to employee participation in decision-making-type programs: only a few workers serve on the firm's safety committee, but our measured per-capita benefits from having a committee apply to all workers.

The same cost-benefit concerns apply to EPFR-type programs: clearly there are (perhaps unanticipated) safety benefits from having employees participate in the financial returns to the firm. But again, these would need to be weighed across the costs. Since 401(k), profit-sharing, and other such programs are probably instituted for reasons unrelated to workplace safety, our findings should tend to confirm their use as a cost-effective HRM practice.

The evidence from Chapter 4 strongly suggests that HRM practices work not so much by changing workers' safety behavior but by improving the safety outcomes of the firm through more efficient use of safety resources. Workers apparently are the lowest-cost provider of information on safety risk and safety improvements, so integrating workers into the strategic planning of the firm and the financial returns of the firm lowers workers' compensation costs. Hence there seem to be two unqualified messages from this research: management involvement with firm safety is important, and workers—even if the firm chooses not to involve them in exactly the types of programs described here—seem to provide useful information on safety outcomes when given the chance. Both worker and management involvement, in the appropriate context, are essential to achieve the optimal level of workplace safety.

IMPLICATIONS FOR WORKERS

So far, most disability policy and research—including policy that affects and research that examines workers' compensation programs—have implicitly assumed that the worker's role is passive, except possibly for claims-reporting moral hazard. It has been assumed that only the employer can handle the formulation and implementation of effective workplace safety programs. This research clarifies the potentially important role that employees play in workplace safety processes. Human resource management practices affect safety on the job: the more employees are involved with strategic safety decisions and with the financial returns of the firm, the lower the workers' compensation costs will be. Moreover, those costs are lowered not only because workers are less likely to file an injury claim, but because their involvement in the safety process changes workplace risk. It is as if the firm, by involving workers, had "hired" cost-effective safety consultants. This should not be too surprising; employees engaged in the production process probably know more about workplace risks than either managers or outside consultants.

IMPLICATIONS FOR WORKERS' COMPENSATION POLICY

One of the advantages of focusing on a sample from a single state is that workers' compensation parameters are held constant. Focusing on a sample of firms from Minnesota means that we didn't have to worry about differences in benefit schedules, waiting periods, or administrative procedures. This was appropriate given our concentration on differences in HRM practices within the workers' compensation system. We have, therefore, nothing to say about how changes in workers' compensation policy or administration may affect the results reported here, except for one obvious implication: any policy or practice in the state's workers' compensation program that prohibits or discourages the types of HRM practices analyzed here would likely be counterproductive.

In other words, in competitive markets, where firms are looking for skilled labor and seeking optimal HRM policies (policies that minimize the sum of the accident costs, as discussed in Chapter 1), the workers' compensation administrator's role should be minimal. Administrators should see that benefits are quickly and fairly paid and that firms bear the appropriate accident costs under the law. Under these conditions, where it is cost-effective, employee participation will occur, and managers will become more involved with the safety processes of their firm. However, not all firms are alike: safety committees may be more effective in some firms than others. Our research does not find that the workers' compensation or public policy administrator should mandate these practices, only that they should not, as a matter of public safety policy, prohibit them from being implemented.

Note

1. This is not strictly correct, of course, since we only measure some of the benefits of HRM practices in our empirical research. We do not measure all of the benefits, perhaps not even the most important benefit (which would be the retention of productive skilled labor on the job). Nor do we measure any of the costs involved in the implementation of these programs. We hope that future research can address these issues.

References

Appel, David, and Philip S. Borba. 1988. "Costs and Prices of Workers' Compensation Insurance." In *Workers' Compensation Insurance Pricing: Current Programs and Proposed Reforms*, Philip S. Borba and David Appel, eds. Huebner International Series on Risk, Insurance, and Economic Security, vol. 7. Boston: Kluwer Academic Publishers, pp. 1–17.

Appelbaum, Eileen, and Rosemary Batt. 1994. *The New American Workplace: Transforming Work Systems in the United States*. Ithaca, NY: ILR Press.

Baril, Raymond, and Diane Berthelette. 2000. "Components and Organizational Determinants of Workplace Interventions Designed to Facilitate Early Return to Work." Études et Recherches, report R-263. Montreal: Institut de Recherche Robert Sauvé en Santé et en Sécurité du Travail (IRSST).

Ben-Ner, Avner, and Derek C. Jones. 1995. "Employee Participation, Ownership, and Productivity: A Theoretical Framework." *Industrial Relations* 34(4): 532–554.

Butler, Richard J., David L. Durbin, and Nurhan M. Helvacian. 1996. "Increasing Claims for Soft Tissue Injuries in Workers' Compensation: Cost Shifting and Moral Hazard." *Journal of Risk and Uncertainty* 13(1): 73–87.

Butler, Richard J., B. Delworth Gardner, and Harold H. Gardner. 1997. "Workers' Compensation Costs When Maximum Benefits Change." *Journal of Risk and Uncertainty* 15(3): 259–269.

Butler, Richard J., Robert P. Hartwig, and Harold H. Gardner. 1997. "HMOs, Moral Hazard and Cost Shifting in Workers' Compensation." *Journal of Health Economics* 16(2): 191–206.

Butler, Richard J., William G. Johnson, and Marjorie L. Baldwin. 1995. "Managing Work Disability: Why First Return to Work Is Not a Measure of Success." *Industrial and Labor Relations Review* 48(3): 452–469.

Butler, Richard J., Yong-Seung Park, and Brian Zaidman. 1998. "Analyzing the Impact of Contingent Work on Workers' Compensation Costs." *Employee Benefits Practices* 1998(4): 1–20.

Butler, Richard J., and John D. Worrall. 1985. "Work Injury Compensation and the Duration of Nonwork Spells." *Economic Journal* 95(379): 714–724.

———. 1991. "Claims Reporting and Risk Bearing Moral Hazard in Workers' Compensation." *Journal of Risk and Insurance* 58(2): 191–204.

Cappelli, Peter, Laurie Bassi, Harry Katz, David Knoke, Paul Osterman, and Michael Useem. 1997. *Change at Work: How American Industry and Workers Are Coping with Corporate Restructuring and What Workers Must Do to Take Charge of Their Own Careers*. New York: Oxford University Press.

Card, David, and Brian P. McCall. 1996. "Is Workers' Compensation Covering

Uninsured Medical Costs? Evidence from the 'Monday Effect.'" *Industrial and Labor Relations Review* 49(4): 690–706.

Centers for Disease Control and Prevention. 2004. National Center for Health Statistics. *Fast Stats A to Z. Work-Related Injury/Occupational Injury.* http://www.cdc.gov/nchs/fastats/osh.htm (accessed November 23, 2004).

Cooper, Cary L., and Judi Marshall. 1976. "Occupational Sources of Stress: A Review of the Literature Relating to Coronary Heart Disease and Mental Ill Health." *Journal of Occupational Psychology* 49(1): 11–28.

Cooper, Cary L., and Michael J. Smith. 1985. *Job Stress and Blue Collar Work.* Chichester, W. Sussex, England: John Wiley & Sons.

Cyert, Richard M., and David C. Mowery, eds. 1988. *The Impact of Technological Change on Employment and Economic Growth.* Cambridge, MA: Ballinger.

Dionne, Georges, and Pierre St-Michel. 1991. "Workers' Compensation and Moral Hazard." *Review of Economics and Statistics* 73(2): 236–244.

Directorate for Information Operations and Reports (DIOR). 2005. *Military Casualty Information.* Washington, DC: DIOR, Statistical Information Analysis Division, Personnel. http://web1.whs.osd.mil/mmid/casualty/castop.htm (accessed April 6, 2005).

Eaton, Adrienne E., and Thomas Nocerino. 2000. "The Effectiveness of Health and Safety Committees: Results of a Survey of Public-Sector Workplaces." *Industrial Relations* 39(2): 265–290.

Fairris, David, and Mark Brenner. 2001. "Workplace Transformation and the Rise in Cumulative Trauma Disorders: Is There a Connection?" *Journal of Labor Research* 22(1): 15–28.

Fenn, Paul T. 1981. "Sickness Duration, Residual Disability, and Income Replacement: An Empirical Analysis." *Economic Journal* 91(361): 158–173.

Fordyce, William E. 1996. "A Psychosocial Analysis of Cumulative Trauma Disorders." In *Beyond Biomechanics: Psychosocial Aspects of Musculoskeletal Disorders in Office Work,* Samuel D. Moon and Steven L. Sauter, eds. London: Taylor & Francis, pp. 207–216.

Fortin, Bernard, and Paul Lanoie. 1992. "Substitution between Unemployment Insurance and Workers' Compensation: An Analysis Applied to the Risk of Workplace Accidents." *Journal of Public Economics* 49(3): 287–312.

Grunberg, Leon, Sarah Moore, and Edward Greenberg. 1996. "The Relationship of Employee Ownership and Participation to Workplace Safety." *Economic and Industrial Democracy* 17(2): 221–241.

Habeck, Rochelle V. 1993. "Achieving Quality and Value in Service to the Workplace." *Work Injury Management* 2(3): 1, 3–5.

Habeck, Rochelle V., H. Allan Hunt, and Brett VanTol. 1998. "Workplace Factors Associated with Preventing and Managing Work Disability." *Rehabilitation Counseling Bulletin* 42(2): 98–143.

Habeck, Rochelle V., Michael J. Leahy, and H. Allan Hunt. 1988. *Disability Prevention and Management and Workers' Compensation Claims*. Final Report to Bureau of Workers' Disability Compensation, Michigan Department of Labor. Kalamazoo, MI: W.E. Upjohn Institute for Employment Research.

Habeck, Rochelle V., Michael J. Leahy, H. Allan Hunt, Fong Chan, and Edward M. Welch. 1991. "Employer Factors Related to Workers' Compensation Claims and Disability Management." *Rehabilitation Counseling Bulletin* 34(3): 210–226.

Habeck, Rochelle V., Susan M. Scully, Brett VanTol, and H. Allan Hunt. 1998. "Successful Employer Strategies for Preventing and Managing Disability." *Rehabilitation Counseling Bulletin* 42(2): 144–161.

Hakala, Leslie Ann. 1994. *Workers' Compensation Insurance at Employee Owned Firms*. Oakland, CA: National Center for Employee Ownership.

Hartwig, Robert P., William J. Kahley, and Tanya E. Restrepo. 1994. "Workers Compensation Loss Ratios and the Business Cycle." *NCCI Digest* 9(2): 1–14.

Hartwig, Robert P., Ronald C. Retterath, Tanya E. Restrepo, and William J. Kahley. 1997. "Workers Compensation and Economic Cycles: A Longitudinal Approach." *Proceedings of the Casualty Actuarial Society* 84(161): 660–700.

House, James. 1981. *Work Stress and Social Support*. Reading, MA: Addison-Wesley.

Hunt, H. Allan, and Rochelle V. Habeck. 1993. "The Michigan Disability Prevention Study: Research Highlights." Upjohn Institute Staff Working Paper no. 93-18. Kalamazoo, MI: W.E. Upjohn Institute for Employment Research.

Hunt, H. Allan, Rochelle V. Habeck, Brett VanTol, and Susan M. Scully. 1993. *Disability Prevention among Michigan Employers:1988–1993*. Upjohn Institute Technical Report no. 93-004. Kalamazoo, MI: W.E. Upjohn Institute for Employment Research.

Karasek, Robert A. 1979. "Job Demands, Job Decision Latitude, and Mental Strain: Implications for Job Redesign." *Administrative Science Quarterly* 24(2): 285–308.

Kasl, Stanislav V. 1978. "Epidemiological Contributions to the Study of Work Stress." In *Stress at Work*, Cary L. Cooper and Roy Payne, eds. Chichester, W. Sussex, England: John Wiley & Sons, pp. 3–48.

Lawler, Edward E. III, Susan Albers Mohrman, and Gerald E. Ledford Jr. 1995. *Creating High Performance Organizations: Practices and Results of*

Employee Involvement and TQM in Fortune 1000 Companies. San Francisco: Jossey-Bass.

Levine, David I., and Laura D'Andrea Tyson. 1990. "Participation, Productivity, and the Firm's Environment." In *Paying for Productivity: A Look at the Evidence,* Alan S. Blinder, ed. Washington, DC: Brookings Institution, pp. 183–243.

Maddala, George S. 1983. *Limited Dependent and Qualitative Variables in Econometrics.* Cambridge: Cambridge University Press.

Majchrzak, Ann. 1988. *The Human Side of Factory Automation: Managerial and Human Resource Strategies for Making Automation Succeed.* San Francisco: Jossey-Bass.

McDonald, James B., and Richard J. Butler. 1990. "Regression Models for Positive Random Variables." *Journal of Econometrics* 43(1–2): 227–251.

Mishra, Aneil K., and Gretchen M. Spreitzer. 1998. "Explaining How Survivors Respond to Downsizing: The Roles of Trust, Empowerment, Justice, and Work Redesign." *Academy of Management Review* 23(3): 567–588.

Moon, Samuel D., and Steven L. Sauter. 1996. *Beyond Biomechanics: Psychosocial Aspects of Musculoskeletal Disorders in Office Work.* London: Taylor & Francis.

Moses, Leon N., and Ian Savage. 1992. "The Effectiveness of Motor Carrier Safety Audits." *Accident Analysis and Prevention* 24(5): 479–496.

———. 1994. "The Effect of Firm Characteristics on Truck Accidents." *Accident Analysis and Prevention* 26(2): 173–179.

Oi, Walter Y. 1974. "On the Economics of Industrial Safety." *Law and Contemporary Problems* 38(4): 669–699.

Park, Yong-Seung. 1997. "Occupational Safety Effects of Employee Participation Plans in Decision-Making and Financial Returns." PhD diss., Carlson School of Management, University of Minnesota.

Rooney, Patrick Michael. 1992. "Employee Ownership and Worker Participation: Effects on Health and Safety." *Economic Letters* 39(3): 323–328.

Shannon, Harry S., Vivienne Walters, Wayne Lewchuk, Jack Richardson, Lea Anne Moran, Ted Haines, and Dave Verma. 1996. "Workplace Organizational Correlates of Lost-Time Accident Rates in Manufacturing." *American Journal of Industrial Medicine* 29(3): 258–268.

Smith, Michael J. 1981. "Occupation Stress: An Overview of Psychosocial Factors." In *Machine Pacing and Occupational Stress: Proceedings of the International Conference, Purdue University, March 1981,* Gavriel Salvendy and Michael J. Smith, eds. London: Taylor & Francis, pp. 13–19.

———. 1987. "Occupational Stress." In *Handbook of Human Factors,* Gavriel Salvendy, ed. New York: John Wiley & Sons, pp. 844–860.

Smith, Michael J., and Pascale Carayon. 1996. "Work Organization, Stress, and Cumulative Trauma Disorders." In *Beyond Biomechanics: Psychosocial Aspects of Musculoskeletal Disorders in Office Work*, Samuel D. Moon and Steven L. Sauter, eds. London: Taylor & Francis, pp. 23–42.

Smith, Michael J., Pascale Carayon, Katherine J. Sanders, Soo-Yee Lim, and David LeGrande. 1992. "Employee Stress and Health Complaints in Jobs with and without Electronic Performance Monitoring." *Applied Ergonomics* 23(1): 17–27.

Smith, Michael J., Barbara G. F. Cohen, Lambert W. Stammerjohn, and Alan Happ. 1981. "An Investigation of Health Complaints and Job Stress in Video Display Operations." *Human Factors* 23(4): 387–400.

Smith, Robert S. 1990. "Mostly on Mondays: Is Workers' Compensation Covering Off-the-Job Injuries?" In *Benefits, Costs and Cycles in Workers' Compensation*, Philip S. Borba and David Appel, eds. Huebner International Series on Risk, Insurance, and Economic Security, vol. 9. Boston: Kluwer Academic Publishers, pp. 115–128.

Taylor, Frederick Winslow. 1947. *Scientific Management: Comprising Shop Management, The Principles of Scientific Management, and Testimony before the Special House Committee*. New York: Harper and Brothers.

Walker, Charles Rumford, and Robert H. Guest. 1952. *The Man on the Assembly Line*. Cambridge, MA: Harvard University Press.

Williams, Cecili Thompson, Virginia P. Reno, and John F. Burton Jr. 2003. *Workers' Compensation: Benefits, Coverage, and Costs, 2001*. Sixth in a series on workers' compensation national data. Washington, DC: National Academy of Social Insurance. http://www.nasi.org/usr_doc/Workers_Comp_Report_2001_Final.pdf (accessed January 27, 2005).

The Authors

Richard J. Butler is the Martha Jane Knowlton Coray University Professor at Brigham Young University in Provo, Utah. He is listed in the third (1999) and fourth (2003) editions of *Who's Who in Economics*, by Edward Elgar Publishing. In insurance, his books and articles have received the following awards: the Kulp-Wright (for best insurance/risk management book published in 1999, *The Economics of Social Insurance and Employee Benefits*, Kluwer Academic Publishers), the Kemper (2002, for best article published in *Risk Management Review* in the previous year), and the Mehr (twice, 2001 and 2004, for articles published 10 years earlier in the *Journal of Risk and Insurance* that have had the largest impact on the field in the decade since). He is on the editorial board of the *Journal of Risk and Insurance*. He served as president of the Risk Theory Society for 1999 and was elected a life member of the society in 2002.

Yong-Seung Park is an assistant professor of human resources and industrial relations at Kyung Hee University's School of Business in Seoul, Korea, where he has taught since 1999. He received his doctorate from the Carlson School of Management at the University of Minnesota in Minneapolis. His research focuses on strategic human resource management, work organization, occupational safety, and labor-management cooperation. His work has been published in the *Journal of Risk and Insurance*, the *Journal of Labor Research*, the *Nordic Journal of Political Economy*, and other social science journals, including Korean-language journals.

Index

The italic letters *f, n,* and *t* following a page number indicate that the subject information of the heading is within a figure, note, or table, respectively, on that page.

Accident costs
  HRM effect on, 1–2, 8–12, 10*t,* 11*t,*
    14, 33*n*1, 89–90
  tradeoffs and, 4–5, 7–8, 8*t*
Accident prevention, achievement of, 1,
  20, 31*t*
Accidents, 4–5, 7–8, 8*t,* 12*n*1, 24
Active Safety Leadership factor, 24,
  33*n*4
Asymmetric information
  HRM and, 2–3, 12
  incentive problem mitigation and,
    6–7, 20

Behavior models
  labor, 2–5, 81, 89
  moral hazard and, 6–11, 20, 61, 86
Bone fractures, 59*n*3
  effect of HRM practices on, 70–71,
    73, 74*t*–75*t,* 77*t,* 78*t*
Business orientation
  employee-owned firms, 18–19, 22,
    23, 27–28
  publicly owned firms, 2, 15–16
  self-insured firms as, 79
Butler-Park study, 27–32, 35–59, 83–90
  aim of, 35
  descriptive statistics, 38*t*–39*t,* 40*t*–
    41*t,* 48*t*–49*t,* 50*t*–51*t,* 52*t*–53*t,*
    57*t,* 66*t*–67*t,* 68*t,* 74*t*–75*t,* 77*t*
  expectations, 27–28, 37
  findings, 28–32, 36–56 (*see also*
    Claim duration; Claim frequency)
  cautions about, 83–87, 90*n*1
  implications of, for
  costs, 56–59
  firms, 87–89
  workers, 89
  workers' compensation policy,
    89–90

California, employee participation in, 23
Claim denials
  claims-reporting moral hazard and,
    64–65, 66*t*–68*t*
  firm's experience rating and, 21
  HRM policies' effect on, 61, 81, 85
  pain and suffering, 62, 82*n*1, 86
Claim duration
  Butler-Park study findings, 44–56,
    48*t*–49*t,* 50*t*–51*t,* 52*t*–53*t*
  employee participation effect on, 18,
    84–85
  HRM practices and, 32, 35, 42, 55
    cost implications of, 56–59, 57*t,*
    85
  unemployment and, 26, 35
Claim frequency
  Butler-Park study findings, 36–44
  HRM practices and, 17–18, 28, 32,
    40*t*–41*t,* 44, 54–55
    cost implications of, 56–59, 57*t*
  Ontario data on, and lost time, 16
  studies on, 24–25
  unemployment and, 26, 42
Claim severity, 18
  duration and, 46, 59*nn*1–2
  quantitative results on, 24, 28, 38*t*
Claims-reporting moral hazard, 9, 19,
  27, 62–63, 69, 80–81
  claim denial and, 64–65, 66*t*–67*t,*
    68*t,* 84, 86
Competition, organizational adaptations
  to, 14, 35–36, 83, 89–90
Concussions and contusions, 59*n*3, 70
  effect of HRM practices on, 70–71,
    73, 74*t*–75*t,* 77*t,* 78*t*
Cuts and lacerations, 59*n*3, 69, 70
  effect of HRM practices on, 70–71,
    73, 74*t*–75*t,* 77*t,* 78*t*

Total quality management (TQM), 18, 30*t*

Trade unions, 15, 39*t*
  member *vs.* nonmember injury frequency, 37, 41*t,* 42
  return to work and, 19–20

Tradeoffs
  accidents *vs.* supervisory costs, 7–8, 8*t*
  accidents *vs.* workers' outputs, 4–5, 11–12, 12*n*1

Trauma injuries, 18, 69

Unemployment
  nonwork spells in Minnesota, 45–46, 50*t*–51*t,* 52*t*–53*t,* 55
  workers' compensation claims and, 26, 35, 42

Unemployment insurance, migrations from, 26, 63

U.S. Occupational Safety and Health Administration (OSHA)
  data from, 30*t,* 31*t,* 40*t*–41*t,* 45
  injury reporting and, 18, 22, 24

Wages, replacements for loss of. *See* Lost wages

Wood-product firms, injury reporting by, 18–19, 22

Workers' compensation insurance
  claims for, 6–7, 16–18, 19, 24, 63, 82*n*2 (*see also* Claim duration; Claim frequency; Claim severity; Claims denied)
  cost of, 1, 5, 5*f,* 7, 26–27, 57, 59–60*n*5, 80–82, 83–87
  downsizing and, 26–32
  firms and, 21, 44, 87–88
  policy for, and Butler-Park study implications, 89–90

Workers' incentives for accident prevention, 1, 80
  alignment of, with management's profit objective, 3, 10–11, 11*t,* 13
  financial returns as, 15–16, 20–23, 89

participatory management practices
  as, 14–20, 30*t*
  problems with, and cost sharing, 6–7

Workers' outputs, 28
  input-output ratios, 7–8, 12*n*3
  tradeoffs in, 4–5, 8, 12*n*1

Workplace environments
  accident prevention and, 1, 2, 4–5, 11–12, 13, 80, 81–82
  data on, as MWSP sections, 30*t,* 31*t*
  employee participation in, and safety, 14–16, 18, 88
  employment uncertainty in, and risk in, and firms selected for study, 29
  workers' compensation costs, 47, 54, 63

About the Institute

The W.E. Upjohn Institute for Employment Research is a nonprofit research organization devoted to finding and promoting solutions to employment-related problems at the national, state, and local levels. It is an activity of the W.E. Upjohn Unemployment Trustee Corporation, which was established in 1932 to administer a fund set aside by the late Dr. W.E. Upjohn, founder of The Upjohn Company, to seek ways to counteract the loss of employment income during economic downturns.

The Institute is funded largely by income from the W.E. Upjohn Unemployment Trust, supplemented by outside grants, contracts, and sales of publications. Activities of the Institute comprise the following elements: 1) a research program conducted by a resident staff of professional social scientists; 2) a competitive grant program, which expands and complements the internal research program by providing financial support to researchers outside the Institute; 3) a publications program, which provides the major vehicle for disseminating the research of staff and grantees, as well as other selected works in the field; and 4) an Employment Management Services division, which manages most of the publicly funded employment and training programs in the local area.

The broad objectives of the Institute's research, grant, and publication programs are to 1) promote scholarship and experimentation on issues of public and private employment and unemployment policy, and 2) make knowledge and scholarship relevant and useful to policymakers in their pursuit of solutions to employment and unemployment problems.

Current areas of concentration for these programs include causes, consequences, and measures to alleviate unemployment; social insurance and income maintenance programs; compensation; workforce quality; work arrangements; family labor issues; labor-management relations; and regional economic development and local labor markets.